Time in Exile

SUNY series, Intersections: Philosophy and Critical Theory

Rodolphe Gasché, editor

Time in Exile

*In Conversation with Heidegger,
Blanchot, and Lispector*

Marcia Sá Cavalcante Schuback

Cover art by Maria Bonomi, *Pêndulo* (1968)

Published by State University of New York Press, Albany

For information, contact State University of New York Press, Albany, NY
www.sunypress.edu

Library of Congress Cataloging-in-Publication Data

Names: Sá Cavalcante Schuback, Marcia, author.
Title: Time in exile : in conversation with Heidegger, Blanchot, and
    Lispector / Marcia Sá Cavalcante Schuback.
Description: Albany : State University of New York Press, 2020. | Series:
    SUNY series, intersections: philosophy and critical theory | Includes
    bibliographical references and index.
Identifiers: LCCN 2019028111 | ISBN 9781438478173 (hardcover : alk.
    paper) | ISBN 9781438478197 (ebook)
Subjects: LCSH: Time. | Exile (Punishment) | Heidegger, Martin, 1889–
    1976. | Blanchot, Maurice. | Lispector, Clarice.
Classification: LCC BD638 .S395 2020 | DDC 115—dc23
LC record available at https://lccn.loc.gov/2019028111

10 9 8 7 6 5 4 3 2 1

To my daughters, Helena and Cecilia,
and also to Andy, my stepdaughter—
all beloved daughters of exiled existence

Remember me, remember me. But ah, forget my fate.

—Henry Purcell, Ur *Dido and Aeneas*

But I want much more than that: I want to find the redemption in today, in right now, in the reality that is being, and not in the promise, I want to find joy in this instant—I want the God in whatever comes out of the roach's belly—even if that, in my former human terms, means the worst, and, in human terms, the infernal.

—Clarice Lispector

Contents

Acknowledgments ix

Introduction 1

Chapter 1
Exile as Postexistential Condition 15
    Times of Excess, Times of Exile 15
    Exile and the Afterness of Existence 19
    The Exile of Memory 22

Chapter 2
The Ecstasy of Time (In Conversation with Heidegger) 31
    The Ecstasy of Time in *Being and Time* 31
    The Ecstasy of Overcoming after *Being and Time* 39
    From a Thought of the Ecstasy of Being to a Listening
        to the Whiling of Being 46

Chapter 3
Time Absent/Time Present (In Discussion with Blanchot) 55
    The Flight of Philosophy into Literature 55
    The Literature of the Step [Not] Beyond 59
    Neither Absence nor Presence—or the Neutral Time of
        the Between 74

Chapter 4
Time Being (Reading Gerundive Time with Clarice Lispector) 83
    Reading Time and the Time of Reading 83
    The Risk of Writing in Gerundive Time 98
    It Is-Being: Or the Neuter Crafting of Life 106

Contents

Chapter 5
Without Conclusion: A Home in Gerundive   113

Notes   127

Works Cited   153

Index   165

Acknowledgments

No one can bear thoughts alone. Indeed, every thought, even when secreted in extreme solitude, is coauthored and shared with many. As such, the activity of thinking is embodied gratitude. *Time in Exile, in Conversation with Heidegger, Blanchot, and Lispector* was written on the basis of the lecture course I held at the Collegium Phenomenologicum at Città di Castelo, in the last week of July 2016, which, under the direction of Alejandro Vallega and Daniela Vallega-Neu, was dedicated to the topic of "Embodied Temporalities: Deep Time, Genealogy and Exile." Invited afterward by Rodolphe Gasché, this lecture course was held again in April 2017, under the auspices of Donato Seminar at the Department of Comparative Literature at the University of Buffalo. Without the illuminating discussions and seminars related to the course, held by an extraordinary faculty during the week at the Collegium, and without the debates at Buffalo University, this lecture course would have never reached the shape of this book. I am profoundly indebted to all who shared their thoughts with me in these very inspiring philosophical environments. First of all, I am grateful to Alejandro A. Vallega and Daniela Vallega-Neu for their invitation and philosophical support to develop this manuscript. My gratitude goes also to Rodolphe Gasché, who invited me to Buffalo and encouraged me to publish the lecture course in his series. I am also very grateful to Krzystof Ziarek, Andrew J. Mitchell, Maria Acosta, and Fanny Söderbäck, who held inspiring seminars related to the course at the Collegium; to Sean Kirkland, Jason Winfree, and Claudia Baracchi, whose lectures contributed to the development of the thoughts I presented; to Ronald Mendoza-de Jesús and Benjamin Brewer and their stimulating comments; to the students who attended the course both in Città and in Buffalo for their questions; and to Ewa Ziarek, David E. Johnson, and Sergey Dolgopolski, who were present at the lectures at Buffalo University and gave me a lot to think about. I would also like to thank the Department of Philosophy at the University of Dundee, where I could once more exchange main thoughts of this book and receive inspiring inputs from A. O. Patrick Levy, Nicholas Davey, Frank Ruda, Ashley Woodward, Oisín Keohane, Dominc Smith, and Michael Lewis. I extend special thanks to Peter Hanly, who, beyond the seminar he also held at the Città, has been a partner in many inspiring conversations about the "gerund" of being. My thanks further go to Jean-Luc Nancy and the discussions we have had in the last years about "being with the without," "the birth of presence," and the "*maintenant.*" My deep gratitude goes to Irina Sandomirskaja and Tora Lane, who read this manuscript with illuminating critical eyes, for the "infinite conversations" that we have shared through the years, and to Benjamin Brewer, who proofread the manuscript and whose attentive comments helped me to find a philosophical language in English for my ideas. Last but not least, I want to thank Italo-Brazilian artist Maria Bonomi, a very close friend of Clarice Lispector, who was gracious enough to allow me to use her work *Pêndulo* on the cover of this book. This work is part of my private collection; no image could be more fitting to a discussion about "time in exile" than these beautiful lines of colors hovering as a pendulum in the event of existence.

Introduction

This book presents thoughts of exile from within, trying to grasp the *experience of time* from within existence in exile. It is a book that aims to do something difficult, namely, to think from the experience of exilic time and not simply about concepts and ideas about time and exile or about personal or collective narratives of exilic experience. It departs from the impact of the time we live in today, but, unlike many other books and discussions of exile, it does not focus on the experience of being outside and displaced. It dedicates its attention rather to the experience of existing inside the outside and to the sensing and senses of time that emerge within exilic experience.

We live today times of the excess of exiles. In the last decades, a huge amount of theoretical literature about exile has been published, and the subject of exile is doubtless a question of increasing political, social, and humanitarian actuality and urgency. Exile has been discussed both empirically and transcendentally, both as human condition and as historical condition and as juridical-political and as psychological-affective issue. It is an old trope in Western culture and has been treated throughout the history of philosophy both implicitly and explicitly. Ontologically, exile has been defined as the movement of all existing things that, as existing, is what comes out of a common ground of being, either nature or God. The "ex," *out of*, that defines exile, is already imprinted in the Latin word *ex*-sistence. But if all existence is a kind of exile from the common ground of nature, of Being or of divine creation, the way existence has been thought and experienced focused mainly on its being-there, its "*-sistere*," "*stans*" or "instances," that is, its standing. The attention to the "ex," to the exilic condition of existence, has been more explicitly emphasized

1

when the question about human existence was brought under scrutiny. Exile, therefore, has been conceived existentially as the proper of human existence qua movement and has been considered what most properly marks the *human* condition. Seneca insisted that human soul is exilic to such an extent that it can never remain where it is, needing to disseminate itself everywhere. And this is true to such an extent that the human soul cannot be exiled from its constitutive exile; thus, nothing in the world is alien to human existence.[1] He explicitly connected human exilic condition to the human struggle for universality. Acknowledging how the human soul is an excess of exile that cannot be exiled from its exilic nature, Seneca summed up ancient views on how the philosophical search for a universal viewpoint and truth presupposes an exile or a flight from the known and owned. For the purpose of exposing this argument, Plutarch, for instance, wrote his famous essay *Peri fugés, De exilio, On Exile*.[2]

These views on exile as constitutive movement of human existence have been present in Greek and Latin traditions since ancient times. In the *Odyssey*, Homer presented the mythological version of what would later define the movement of human existence as the longing for the freedom of truth and the truth of freedom: in this version, human existence is an odyssey, the movement of departing from the known, adventurously traversing the unknown, and coming back to the known transformed by the unknown. This *Odyssey*like trajectory has been used for centuries to define both the movement of consciousness and of exile, reaching modern times in Hegel's attempts to describe the phenomenology of the spirit and Schelling's views on the "Odyssey of the Spirit."[3] Thus, for both, the trajectory of the Spirit is essentially exilic. It is found as well in the description of the platonic cave as the very structure of philosophical *paideia*.[4] With slight variations, this exilic scheme is operative in the neo-Platonic heritage of Christian tradition through which human existence is described as the movement from exiting [*exitus*] God, living a worldly existence in dispersion and disquiet [*cura*] and searching for return [*reditus*] to divine unity through grace.[5] Moreover, exile marks even more emphatically the Jewish tradition insofar as it defines not only the fate of a people but also the very meaning of being a people. This fate and meaning are anchored by the heavy accent of expulsion and persecution of exile, which renders discourses on the "wandering Jew" both ambiguous and problematic.

Franz Rosenzweig insisted that the Jewish people can be called the "eternal people," "the people that becomes the people, as in the dawn of its earliest times so later again in the bright light of history, in an exile."[6] In regard to the Jewish religious tradition, to its history and culture, exile is what guarantees that the Jewish people "is a people only through the people."[7] Defining the condition from which the Jewish people can be a people and further can understand itself as eternal life, exile marks not only the way human existence exists but in which sense it can exist as a people without defining its being on the basis of its belonging to a territory.

Until modernity and mainly through the heritage of ancient Greek, Latin, and Christian tradition, the exilic scheme of human existence, proposed by the philosophical tradition, was marked by the promise of a return to home, to Nature, to God. It was marked by nostalgia. We could say more simply that, before modernity, human existence was philosophically conceived of as existence in exile but an exile that keeps promising the return. Modernity understands itself as historical exile. In modernity, exile becomes itself a historical condition. Modern existence is a no-longer existence, no longer "Greek," no longer "religious," no longer bound to tradition and authority, as Kant announced, but an existence continuously breaking with its own past. It is existence in renaissance, in reform, in revolution. In this sense, modernity is grounded in an exile without return, thus every "return" described in modern terms, is return to an invented, constructed and forged beginning or origin. Modern promises are other than those of the Ancients: they are promises of revolution, of grounding what had never before been had or seen, either in the encounter with the New World or in the forging of new forms for the world. At the same time that philosophy wants and longs for a home everywhere, recalling Novalis's famous quote—"Philosophy is homesickness, the urge to be at home everywhere" (Die Philosophie ist eigentlich Heimweh, Trieb, überall zu Hause zu sein)[8]—exile is romanticized as the necessary suffering condition of creation and conquest. Exile without return, as the common English saying goes—"You can't go home again"[9]— is the structure of modern concepts of *Bildung*. In postmodernity, though, exile becomes a condition of the world. As a condition of the world, exile knows the extreme form of exile, not only without return but also without departure. It is exile without departure and also without arrival. This appears very clearly in the second generation of histories of exile, in the children of exile, who have never known the "before" the exile and continue to experience the denial of an "after." In this extreme form of exile without departure and without arrival, without promises of return or arrival, exiled existence becomes existence haunted by the violence of the extreme, what includes, of course, extreme forms of violence. Exiled existence is existence at the edge, at the frontier, continuously touching the frontiers and edges of existence.

According to this sketchy history of the expansion of the concept of exile from a juridical-political concept to ontological, historical, and epochal concepts, in which exile defines not only a *conditio humana*, but also a *conditio historica* and further a *conditio mundana*, two dimensions are continuously intertwined: the concrete juridical-political experience of existing *in* exile and the existential-historical-ontological understanding of existence *as* exile. In all these formulations, exile is understood on the basis of the ecstatic exilic scheme of movement, grasped as an ecstatic change from-to. Described as the torment of loss or as a resource for creation, exile has been grasped for centuries as the narrative of a cut, of an interruption or of a caesura that separates in more or less absolute ways a before from an after, an origin from a destination. Existence in exile, then, has been understood as existence in the cut that separates and interrupts the continuity of a before and an after. Thus, what is interrupted here is the continuity of time itself. This continuity is interrupted because the no-longer-being of the past and the not-yet-being of the future not only remain always present but become even more present than the present, not solely in the sense that the present would dim or fade away under the overexposure of past memories and future expectations. The point to be made here is rather that exile interrupts any experience of time as continuous succession of before and after, the very measure of the movement of this flow, precisely because, in exile, existence is suspended in the *between*. As such, it can be said that exile is the experience of the *epokhé* of existence. It is a countertime in the time of existence and in existing time. Countertime means here *both* another time in the order of time *and* something other than the order of time, as Werner Hamacher shows quite clearly in one of his last essays published in English.[10] As existence suspended in the between, exile can indeed be considered untimely existence in time.

But how to conceive of and find an adequate formulation to describe the experience of exile from within, that is, from exile while exiling, indeed an experience from the while while whiling, so to speak? In question is no longer "time," neither a tense, nor a "voice"— not even the middle voice—but a verbal tension, which is the proper experience of what grammarians call the "gerund."

The term "gerund" derives from the Latin verb *gerere*, which means to carry on, bear, bring forth.[11] The fundamental meaning is of an on-going action, without provenance or destiny, without a beginning or end. Grammarians define the gerund as a verbal noun because it has the property of acting both as a verb and as a noun, being a kind of "halfway" between both.[12] It is close to the present participle, and traditional grammar sometimes considered them as synonymous. Besides the gerund, the Latin language and grammar also knew another form of verbal noun, very close to the gerund, called "gerundive." Those forms are also called nominal forms of the verb or verb nouns insofar as they do not carry any mark of temporal or modal flexion, assuming characteristics of a noun even though they are not nouns. They are, in this sense, also very close to the infinitive. Gerundive forms reject articles and work, so to speak, as nouns against substantivization. The distinction between the gerund and the gerundive in Latin is not easy to explain but can be described as following: the gerund is a verbal noun, always *active* in force, having the infinitive in the nominative case, and the other cases formed with a -nd to the present stem of the verb. The gerundive has more the function of a verbal adjective, passive in force, formed by *–ndus* and related forms added to the stem of the verb.[13] According to French grammar, verbal nouns ending with *–ant* are called *geróndif*. Grammarians have great difficulty in accounting for this difference, above all because the gerundive disappeared in Latin, and some modern languages tend to use infinitive forms to render its meaning.[14] Some grammarians want to read in the gerundive a kind of *participium necessitatis*, in which a mandatory meaning and a futural sense of "having to be" seem to be implied. In contemporary English grammar, the gerundive is difficult to discern, but a possible way of rendering it would be with "to be- done, read, said," and so on. In languages such as Brazilian Portuguese, however, where gerundive forms are abundant and appear in multiple uses, the main sense of these forms is the expression of a continuous action, without any idea of fulfillment, achievement, or end, being thus also understood as a nonfinal verb. Gerundive forms are fundamentally performatives, describing an action while acting. I will use both expressions "gerund" and "gerundive" (time or temporality) to express the on-going and whiling in the between and not the grammar uses of it as verbal noun.

What I am trying to stress here differs from some recent reflections[15] on the gerund and gerundive that can be read for instance in Samuel Weber's discussions on "theatricality as medium"[16] and in the beautiful thoughts by Pascal Quignard on "the image that is lacking in our days."[17] Paying attention to the gerund and present participle as the grammatical hallmark of a certain meaning of "theatricality," Samuel Weber considered these modes to be the ones in which presence is suspended, letting appear an interval that links and separates what is presented and the presentation, constituted by a series of repetitions, which are modes of disjunctive "goings-on" that anticipate the future remembering the past.[18] Describing the gerund and the present participle in these terms, Weber still "reads" it from the viewpoint of a temporal sequence; even if the main focus is the way the present gives itself in this sequence. For Pascal Quignard, the question is not about the anticipation of the future remembering the past but of "being before a to be done" [étant devant être fait], insofar as the focus of his musings on the gerundive is the Latin mural painting, which he recognizes as the painting of the image that is lacking in the image. For him the gerundive is always saying "devant," which in French can mean "before" in the sense of in front of, but also "having to," and, last but not least, is a form in which the gerundive form -ant is always present. The act of seeing an image is therefore intrinsically gerundive, car it is a "devant devant devant,"[19] a difficult phrase in his short chapter that maybe could be rendered with "having to be-being in front of, or ahead of itself." The main sense Quignard acknowledges in the gerundive is the imminence, the "ambush" of being, the "scopic instant" as he also calls it. But also here it is the future and the infinite that seem to define the gerund and its gerundive tension.

Even if these aspects definitely can be attributed to the "gerundive" and the present participle mode of being, what seems to me decisive is, nonetheless and above all, the type of movement that constitutes the verbal tension in a between and a meanwhile, a movement enigmatically without movement, that can only be conceived of when the scheme of movement as a change from-to is somehow forgotten. This is in my view what defines time in exile.

The main thesis of this book is that the meaning of exile is to be grasped from the gerundive sense of present, which differs and exceeds the present and its modes of presence. As such, the meaning of exile is not conceived from the ecstatic scheme of movement as change from-to, where a view towards the before and after, with their emphasis on mourning and memories, on broken utopias and futures, and on losses and frustrations, directs attention away from the core of exile—namely, from the suspended existence in the between and in the meanwhile.

Fixated on the ecstatic scheme of movement as a change from-to, and on the consequent focus on displacement, discourses and philosophies of exile become inattentive to what Walter Benjamin once called "the perceptive now" and that we could extend in terms of "the perceptive now of exile." They remain indifferent to how exile gives itself to view from *within*, that is, from its experience as an immense struggle for presencing. Inattentive to how existence in exile is suspended in the between and in the meanwhile, theories and discourses on exile become blind as bats facing the light of a simple truth, to use an image by Aristotle in the *Metaphysics*.[20] The simple truth of exile—that is, how it shows itself from within itself, from its experience—lies in the between-existence it exposes one to, neither here nor there, neither in the before nor in the after, not even exile nor asylum, but a disquieting interstice, an existence at the edge, at the frontier of existence.[21] Time in exile is the time of existing-between and in the meanwhile, time in which a sense of present more present than the present exposes itself. What emerges here is a nearness closer than proximity, a fragile groundlessness, insofar as it can find a ground neither in the past nor in the future. Neither before not after, thus neither-nor defines exiled existence as a *neuter*; neuter, which in Latin means "neither . . . nor," but never "neutral," thus in exile nothing can remain neutral or indifferent. What marks the neuter—neither-nor—of exile is however not so much what is negated—as for instance the here or there, the before or after—but the hyphen and the mode of its presence. In the experience of being neither-nor, one experiences being not as nothing but as is-being. Existence in exile is, above all, existence exposed to the uncanny is-being, to the bare "is-existing"; an odd expression in English that aims to express the gerundive mode of being in exile. It is existence unable to rely upon what it was or on what it can, could, or would be, having nothing to rely upon except the is-existing. It is existence in gerundive. In exile, existence is complete insecurity; thus, the only thing that remains is not even language, as Hannah Arendt affirmed,[22] but merely the "is-existing." This bare, unsheltered, exposed, and exposing "is-existing" is indeed the only place and time of exiled and exilic existence, a place without a place, and a time without time, a groundless ground to exist. Rather than a question of space and time, existence in exile poses the question of the between and the meanwhile. Existence in exile is indeed existence in disquiet. Every existence in exile is a kind of *Book of Disquiet*, to recall the title of the Portuguese poet Fernando Pessoa. The is-existing, upon which existence in exile is sustained, is indeed without exit; thus, it is not possible to escape from this placeless place and timeless time. Is-existing shows itself as the moving placeless place and timeless time from which no one can be moved away. Is-existing, the gerund of being, means an excess of nearness, a too-excessive nearness that can hardly be borne or carried out; if images of the past and of the future appear to be so emphatic in exile, it is because the groundless is-existing is too unbearable, too overwhelming to be carried out, like a shivering bird in the hand. This trembling nearness of the is-existing can be called *presencing*. Indeed, shivering *presencing* is what human reality can hardly bear and stand.

To pursue this thought of exilic existence as existence suspended in the between and meanwhile, as the thought of the is-being and its gerundive temporality, I propose a reading conversation with three authors that provide elements for a thought of gerundive time, however, in quite distinct paths. The three authors are Martin Heidegger, Maurice Blanchot, and Clarice Lispector. If Heidegger and Blanchot can be read as thinking and writing time and being in the excess of a withdrawal, Clarice (known simply as Clarice in Brazil, without any need to add a family name) can be read in turn as a writing of time being, as gerundive writing. What must be strongly emphasized is that the purpose of this book is to establish a conversation *with* these three authors about the exilic experience of time and not to give an account of their thoughts or even less of a conversation among them. My focus is *not* on how and to what extent Heidegger, Blanchot, and Clarice are interconnected but on their thoughts on the problems I proposing discuss. Of course, there are strong connections among these authors, but I aim to bring their voices as partners in my search rather than to search for their connections and disconnections regarding a problem. The constellation of these authors concerning the question about the sense of time in exile may appear odd. Neither Heidegger nor Blanchot is an author of exile, in the sense literary research on exile has rendered canonic.[23] In many aspects, they are the opposite. Heidegger is an author of rootedness and never speaks of exile. Blanchot, who does speak of exile, is, nonetheless, an author of not-departing, not in the trivial sense that he has not left France or his language but for a writing that goes continuously back and forth, affirming and negating each affirmation and negation at the same time, assuming the task of not-departing as the only way to overcome fate and language. They are, nonetheless, decisive authors in regard to a thought of time that breaks with the rigid chain of chronology and presence and that addresses the essence of time in terms of absence and withdrawal, of ecstasy, and of the event. The thoughts of Heidegger and Blanchot have indeed provided an important basis to contemporary discussions about the meaning of exile and the writing of exile, in which the sense of exile as excess of loss and withdrawal of presence have been emphasized. Clarice Lispector, whose work has become the object of increasing interest and study in the last years, has a different position in this constellation. Despite being herself a child of Jewish exile, she never made exile one of her literary tropes, developing a literature that can be most precisely described as the writing of gerundive time. If Heidegger and Blanchot can be read as thinking and writing time and being in the excess of a withdrawal, Clarice (known simply as Clarice in Brazil, without any need to add a family name) can be read in turn as a writing of time being, as gerundive writing.

These three authors share several figures of exile, but in very different modes. They have the *neutre*, the neither-nor, which is a mode to formulate the between and meanwhile at stake in exile, as explicit figures of their thought and writing, but in very different senses. The three think throughout the "it," "*Es*" in Heidegger, the "*il*" in Blanchot, "*it*" written and pronounced in English in Clarice who lets this foreign pronoun enter into and encroach upon the Portuguese language. Three ways of thinking stepping: in Heidegger, stepping beyond by stepping back, *der Schritt zurück*; in Blanchot, stepping not-beyond, *le pas au-delà*; and in Clarice, the steps in the being-on and in and never beyond. Three ways of thinking the way: Heidegger's thoughts on the way conducting nowhere, *Holzweg*; Blanchot's thoughts on the way of the no way, and Clarice's thoughts on the sway and swing. Three thoughts of the excess, with different thoughts of the excess of being, and three approaches to presencing: Heidegger almost approaching presencing as time in gerundive; Blanchot avoiding every thought of presencing but seizing its decisive absenting; and the thought and writing experience of time being in the gerundive in Clarice's literature, in which the excess is itself exceeded. If Heidegger is a philosopher that can be considered the "philosopher of philosophy," Blanchot is more of a theoretical writer, in between philosophy and literature, and Clarice is a writer writing all the time the coming to writing, near the "wild heart" of the is-being. Even if the book will not focus on questions of gender, one should not forget that two men and a woman form this reading constellation. Moreover, this constellation gathers a German and a Frenchman, with the historical and political implications of these citizenships and of their political positions, and a Jewish Ukrainian who emigrated to Brazil as a very young child before World War II; a philosopher searching to overcome philosophy, a theoretical writer trying to overcome both theory and literature; and a writer not trying to overcome anything. Two family and tradition names—Heidegger and Blanchot—one given name, Clarice. And, last but not least, they present very difficult modes of writing, for they are all three excessive and very demanding writings, challenging readings, writings that are already readings, near to saturation, deeply performative and "afformative" (Hamacher),[24] rendering impossible interpretative methods and the control of memory, insofar as all the three writings tend to erase themselves through the very writing, and the reading must learn to read as the movement of an approaching, and thereby to be totally disarmed, knowing that it can only approach and never be close. Represented are three experiences of thinking writing that demand we think and speak in the difficult language of translation; thus, even if Clarice is for me literature in my mother tongue, she writes with a foreignness as if it were in translation.

An underlying thread that brings Heidegger, Blanchot, and Clarice together in relation to the question of the book is the relation between philosophy and literature, an explicit question for both Heidegger and Blanchot and an implicit intertwining in the work of Clarice. In her work, this question appears precisely when it no longer is at issue. What brings together philosophy and literature is not in the first instance the relation between the poetical and the conceptual or abstract (Valéry),[25] but the event of thought in language and of language in thought. Because one is so intrinsically connected to the other, to the extent that one should rather say that one is already the other, the logic of causes and effects, the logic of "therefore" is not capable to account for it ("I think therefore I speak" or "I speak therefore I think," are unsuitable formulations), and the temporal sequence of a first followed by a second appears untenable. Heidegger, Blanchot, and Clarice, are authors for whom the awareness of the enigmatic intertwining of thought and language appears in the experience of thinking in language, of thinking being language and language being thinking. These three authors, in very different ways, are not only attentive to but somehow obsessed with the *experience of thinking in language*, with the event and happening of thought and language, and not only with a reflection about the relation between thought and language. To be thinking, to be saying, to be writing, to be reading—these experiences are nothing but experiences of gerundive time, of thinking of thinking, saying, writing and reading *while* thinking, saying, writing and reading. In the very experience of the event of thought and language, indeed of thinking in language and of the language of thinking, elements for thinking the gerundive mode of time, experienced from within exiled existence, can be found. Because these authors are, at different levels and in various degrees of intensity, approaching and immersed in the difficult attention to thinking, saying, writing, and reading the "while" [I am] thinking, saying, writing, and reading, they can be considered thinkers of the experience of gerundive time, thinkers of experience from within.

This book is structured in the following way. The introduction presents a discussion about the general aim of the book, followed by introductory remarks about the meaning of exile. The second chapter lays out thoughts about the times we live in today, taken as times of excess and of exile. It frames the hermeneutical situation of our today and the urgency for thinking otherwise the experience of exile as a matter of gerundive temporality. It deals with the need to reformulate basic presuppositions that orient most theories and literature of exile, namely, that exile is structured by the change from a place to another, that it is a cut and interruption of the continuity of time, and that memory is anchored in the past. Questioning these presuppositions, the chapter opens up the necessity to engage with gerundive temporality, the temporal movements of the between and of the meanwhile, in the search for a thought of exile from within. In the third chapter, a conversation with Heidegger aims to show how his thoughts on the ecstasy of time come to a point that almost touches the question of the gerundive mode of time. This happens in Heidegger's late discussions about presencing, whiling, and abiding. It could be said that Heidegger is on the verge of thinking in the gerundive. Investigating why he did not reach this thought, it appears that what prevents him for thinking in gerundive is the thought of the withdrawal and of overcoming, always operative in his ideas of ecstatic time and being. The fourth chapter engages in a discussion with Maurice Blanchot, following out his concerns with the figure of withdrawing and of the outside as tension between time absent and time present. If Blanchot explicitly attempts to avoid the Heideggerian path of Being, he did keep and even accentuated the thought of the withdrawal and of the excess, as the only mode of presence. Also in Blanchot, it is possible to observe how he comes close to a thought of gerundive time precisely in the reluctant way he addresses the problem of the presencing of time and affirms presence as withdrawal. It is, however, the way he connects the figure of the outside with the dynamic of withdrawing and absenting that distances him from gerundive time. The fifth chapter presents a reading *with* rather than *of* Clarice Lispector, particularly with her novels *The Passion according to G. H.* and *Água Viva*, which shows how Clarice is a writer of and in the gerundive mode of time. For what the former thinkers almost touch upon—namely, a thinking-saying-writing-reading experience of time in the gerundive—is in fact the quintessential element of Clarice's work. At the end, in lieu of a conclusion, a discussion about what it might mean to dwell in the between and meanwhile of exile, and how one might formulate the sense for a home in gerundive.

Because the gerundive mode of time in exile is in question, the experience of the is-being while it-is-being, this reading will thus attempt what I propose to call an "approaching reading" of some passages of these authors of extreme per-a-formative writing. By "approaching reading," I understand a reading that is somehow closer than close reading in the sense that it aims to follow the formless movement of how the text is writing down how it is being read.[26] It is a reading in a certain sense more similar to how a drawing is "read," that is, seen, following its being-drawn by the drawn lines. Approaching reading is a double reading, a reading to the extent that one designates and to the extent that one cannot designate, a reading "searching and not finding that what I did not know was born, and which I instantly recognize," to quote Clarice. "Approaching reading" should be understood here as a reading attentive to the approaching of a thought in language and of the language of a thought. It is a reading immersed in the attention to the approaching of thoughts in words and words in thoughts. The approaching reading proposed in this book is neither a "close reading" nor a "comparative reading" of Heidegger, Blanchot, and Clarice. It is a way—a "method," we could say—to approach the approaching of a coming to thought, to language, to writing, the approaching which expresses the gerundive mode of time of the "is-being." Because the main thesis of the book is that in exile, time is experienced as gerundive and that gerundive time is not a present tense but the tensioned meanwhile, itself back and forth being neither back nor forth—what the awkward expressions in English "is-being" and "is-existing" aim to call attention to—it can be said that the whole book is about approaching the meaning of an approaching, of the imminent, of the "about to" happen and be, that so deeply and painfully marks existence in exile and the experience of time in exile.

Chapter 1

Exile as Postexistential Condition

Was geschieht, ist Abschied

—Werner Hamacher

Times of Excess, Times of Exile

In which time do we live today?[1] This question is a concern to every-one. This question pierces the world. It became common sense to question in academic discourses the use of the pronoun "we." Which "we"? From which viewpoint can we say "we"?—these questions have of course their legitimacy and necessity, but, in asking "In which time do *we* live today?" what is being addressed is the question about the awareness of the world today. Saying "the time we live today," the "we" indicates how the today is experienced as a common moment of history. Whatsoever "we" pronounced today experiences the world of today as what overwhelms differences and particularities, indeed as the "global" world. Global is the experience of a world, on the one hand, too heavy of world and, on the other, of a world without world. Without any attempt to give a definite answer to the question "in which time do we live today?," it can at least be said that our time is a tired time, a tired time of the "too much": too many things, too much information and disinformation, too much dispersion, too much despair, too much ambiguity, too much velocity, too much for few, too little for many, too much too much, too much too little. Indeed, our time is a time of excess: excess of the capitalism of excesses, of the capitalism of misery, of war, of segregation, and of ambiguity. Looking at the *Oxford English Dictionary* for the origin of the word "tired," we find its oldest Celtic form *tirian*, also spelled as *teorian*, which, in an unscientific and associative manner, brings us to the Greek word *theoria*, theory. Tired of theories, the world is as well. Nikolai Chernyshevsky's question *What Is to Be done?*, posed by Lenin, is screaming everywhere, from the most extreme to the most conservative positions, in the West and in the East, in the South and in the North, and in their many middles. The misery of theories haunts us even in attempts to develop sophisticated theories of contemporary misery. The times promise no rest. But this promise of no rest should be taken seriously, and the face of impotence that frightens the world should be confronted with the force of existence itself, as Jean-Luc Nancy appeals us to do when discussing anew the question about what to do.[2] "We," at least we readers of the world, should take this promise seriously and rest in this lack of rest, remaining in this arrest of the times. Rather than looking for a way to escape the exhaustion of having no rest, of having no place or time to rest in a time of excess, we should dwell the question about this time of excess and try to listen attentively to its "cardiography," that is, to how the passions of the "wild heart" of time beats in these times of excess.

One can hardly deny that the times are now times of excess. Excess, however, does not mean merely too much. It means a "too much" that goes beyond all measures. It also means outrage. Indeed, the primary meaning of the word "excess" and of the verb "to exceed" is precisely the violence of going beyond limits. To go beyond limits is also the meaning of another verb that also comes from Latin, namely, "to transcend." Addressing our time as a time of excess is, however, not the same as addressing it as a time of transcendence. As a matter of fact, it is the other way around; thus, these times of excess, that we recognize as "our" time, the hour in which different "ours" attest and bear testimony of the excess and lack of the world in the world, are times of total immanence, times without any horizon of a beyond, times without a way out, that are not only times without an outside but also without a without. Thus, to transcend—either a state of mind, a situation, or a condition—means to get out of it and, hence, to get to be without it somehow. Trance states are states in which one is out of oneself, without oneself, or to be more precise, states in which one is with the without of oneself.[3] The anxiety of these times of excess is the anxiety about the impossibility of even glimpsing a way out, about the disappearance of the horizon of transcendence. One could, therefore, speak about times of excess without transcendence, indeed of times of *un-transcendental excess*. Speaking in this way, it is acknowledged that this excess of the times, of "our" time, has a lack, and thereby that this excess of "our" times is *lacking*, lacking something even in its excess. Indeed, and despite the strangeness of this affirmation, something is lacking in these times of excessive excesses. What is lacking is the lack. However, "our" times can be considered times of excess not only for being experienced as times without a way out, without a without, of times lacking a lack, but above all for being times in which the excess of being appears "at the flower of the skin," to translate literally a common expression in French and Portuguese, *à fleur de peau, à flor da pele,* used to express how an overwhelming experience comes to be not only under our skin but to be *our own* skin. In which sense do the times, "our" time, experience "in the flower of every skin" the excess of being? The excess of being appears when to be means to become whatsoever, whenever and wherever, for the sake of being capable of being used, abused, misused by who- and whatsoever, whenever and wherever. This capability to be used is often called today "empowerment." In times of excess, the excess of being appears when being means nothing for being whatsoever. Everything must lose ontological determination for the purpose of receiving any ontological determination whatsoever, depending on the "demand." In this sense, excess of being means continuous dis-ontologization, which renders possible continuous re-ontologization. The excess of being in times of excess does not mean simply that a certain meaning of being becomes empty, as Husserl indicated when discussing the meaning of technique in the *Crisis of European Sciences*[4] nor that a certain meaning of being becomes omnipresent, as Herbert Marcuse claimed in his discussions of the *One Dimensional Man.*[5] The excess of being in times of excess means rather the emptiness of omnipresence, the entropic dynamics of the more *beings,* the less *being.*

Excess without transcendence shows itself as the times of a suspension of movement when movement cannot stop moving. If "our" time identifies itself as the time of the excess of the capitalism of excesses, it is not merely because capitalism relies on movement measured by time, but because it relies on untiring, constant, and restless movement, that is, on a movement that cannot stop moving, a movement that, while moving everything, does not move itself. Excess is not a state beyond but the dynamics of beyond, indeed a dynamics of movement insofar as every movement implicates a beyond; that excess is a dynamics of moving beyond is something that was already grasped by Greek philosophy, and more specifically by Aristotle and his concept of the "unmoved mover," οὐ κινούμενον κινεῖ,[6] and that Marx recognized as the thermo-dynamics or, as we could say today, the "global warming" of capitalism. Thus, what is capitalism if not the mover of everything that remains itself unmoved? Indeed, in capitalism, capital is the untiring movement of everything, so that everything can be anything and anything can be everything precisely on account of its being nothing. Indeed, capital rests on the assumption that being is nothing, nothing that could not be something else and otherwise.[7] Formulated in these terms, the excess is not merely the too much but a strange reversal of itself. It reverts itself becoming terror, reverting all creative force against creation, as the modern history of revolutions has showed. But it also reverts itself, rendering movement unmoved. Not a reversal in the sense that a countermovement would force movement in a certain direction or would resist movement; it is a reversal of movement *inside* movement, a point or layer of intensity in which the movement moves against itself. In its utmost intensity, movement does not cease to move but remains moving, and hence moving without moving the movement itself. In the excess of movement, movement is reversed into its own suspension.

The difficulty of seizing such an excess of movement that suspends movement lies in the deep-rooted idea of movement as change from one place to another, from one time to another, from one state to another—in short, in the idea of movement as a change from-to, which implies a thought of the beyond. This "from-to" defines what can be described, following Lacoue-Labarthe's discussions on mimesis,[8] as the "arche-teleological" structure of movement, which has been moving centuries of theories on practices and of practices of theory. "From," is said in Greek with "ek/ex" and "to," "onto" with "eis." The Greek word that condenses both and expresses this arche-teleological structure of movement as change from-to, ex-eis, is "ecstasy." That is why in the *Physics*, Aristotle defines every change, *metabolé*, as what is by nature ecstatic: *Metabolé dé pása fúsei ekstatikon*.[9] In this Aristotelian framework, "ecstasy" does not deny the arche-teleological structure of movement in which movement is defined as dislocation *from* a place *to* another but determines this structure. If in its current meaning ecstasy names what disturbs or subverts the teleology of movement, it is because in ecstasy the from-to reveals itself as a structure beyond the places from where and to where the movement moves. As change from-to, movement is in its nature ecstatic, this is the Aristotelian position, because moving things exit their original and/or natural positions. What allows Aristotle to describe movement as change and change as in its nature ecstatic is the viewpoint from which movement is "read" from things exiting positions. It is thereby seized upon as way out into the beyond, and as such as an expulsion, what in Greek is said with the word *exodos* and in Latin with *exilium*. As the arche-teleological structure of a from-to, movement is in itself ecstatic, exodic, exilic, insofar as it shows an exit towards. Therefore, assuming that what defines exile is ecstasy, that is, a moving beyond frontiers, a getting *out* of, a going *away from,* an exit, it can be said that it is exile that defines movement rather than movement that defines exile. If movement is ecstatic, exodic, exilic, how are we then to grasp the excess of movement? This would be an excess of ecstasy, of exodus, of exile, in which there appears to be no exit from continuous exiting. What is generated by this exit without exit, by this continuous exit or transformation is the status quo.[10] Thus, in order to be continuously exiting or transforming, transformation cannot transform itself. The Greek word for status quo is *stasis*, a word that also means civil war, the war of the self against itself, of the same against the same.[11] The excess of ecstasy, of exodus, of exile, of excess generates *stasis*; in excess, ecstasy becomes stasis, and the ecstatic is rendered static. At this point, it becomes possible to ask if the figure of ecstasy and its excess, as the only imagining that remains of transcending a state without exit, is still the proper figure to think "our" times of excess and exile, indeed of the excess of exile.

Exile and the Afterness of Existence

The connections among exile, ecstasy, and excess that orient most literature and philosophies of exile are grounded on the understanding of exile as the condition of existing *after* a cut or interruption of a world. With the cut and interruption of existence that marks exilic condition, the past and the future no longer appear as what follow naturally one from the other but as a cut past and a cut future, and as such as estranged past and future. In exile, the natural feeling that one belongs to a past and that a future belongs to such a past becomes deeply questionable. The whole existence is pervaded by questions such as "Where do we come from? Where are we? Where are we going?"

These questions, which entitle a beautiful painting by Paul Gauguin, are repeated today by millions of tired and hungry mouths without recourse. These are questions shared by more and more people, but they are also questions that more and more divide the world into several forms and registers of "we" and "they." A discursive wall is being built around these questions that do not only divide "the own" and the "foreign," but also the owness of strangers and the stranger's owness and, further still, the owness of the own and the strangers' strangerness. In countless refugees' stories, the mythological figure of the deluge is repeated, in different languages and dialects, yet with an important variation. The variation is mainly that one is no longer expelled from paradise but from war and poverty, and this without having committed any fault, without any guilt. What is repeated is the constant exile and the search for asylum, which coincides today with searching for entry into a system of debt and guilt. (Thus, what is the West today if not a global system of debt and guilt?) Exile describes a disruption that does not only disrupt all of existence and its concrete conditions but that disrupts, above all, the sense of what it means to exist. It produces a caesura within existence by which not only the past and the future have to be reinterpreted and reevaluated, but according to which even the actual perceptions of time and space demand to be rephrased and reintuited. In exile, existence is not only the exit from previous existence or from nonexistence, as the Latin word *ex-sistere* expresses, but is existence after having existed. It is after-existence. Exile puts existence in a condition of postcondition.

In Latin, the adverb "after" is *post*. Exile can be considered the condition of postexistence. Rather than a postmodern condition, the world experiences today a *postexistential condition*. This expression differs from the recurrent tropes of postisms within humanities: postmodernism (Lyotard),[12] posthumanism (Foucault),[13] postcolonialism (Young, Spivak, Bhabha),[14] postcommunism (Boris Groys),[15] etc. It differs because it aims to describe the sense of existing in a postcondition and not only to summarize diverse narratives and nonsystematic knowledge conducted within humanities and brought forward by different kinds of art.[16] In the expression "postexistence," what is central is not so much the question of what it means to exist after a certain event—after a separation, a cut, a catastrophe, a disaster, or after a trauma—but rather the sense of existing *as* "afterness" itself.

The expression "afterness" translates the German word *Nachheit* and was coined as a culture-critic concept by the German literary scholar Gerhard Richter.[17] In German, *Nachheit* not only means afterness but also implies nearness [*Nah-heit*], since in German *nach*—after—and *nah*—near—almost sound the same. They sound as a lingering note that through a smooth glissando passes from one meaning to the other. This lingering note plays an important part in the concept of *Nach-/nah-heit* since, in focus here, is on the one hand the demand for a listening, and on the other the acknowledgment that nearness is a decisive experience in a postcondition. Afterness is a condition that remains constantly separated but also always near to what it has been separated from without return. Afterness is the condition of a disquieted and stressed nearness to a far away. In the book called "*Afterness*," in Richter's own translation of the German word *Nach/nahheit*, figures of following in modern thought and aesthetics are investigated. His thesis is that modernity *today* is to be understood as a time that can only perceive itself, that is, its own time as an after modified by this after, a thought developed in the book under the inspiration of Jean-François Lyotard's statement that "after philosophy comes philosophy but changed by this after."[18] In order to understand the meaning of this modification, which does not let itself be thought in terms either of an invention or of a revolution, the focus must turn toward the after as a specific movement and dynamic. The question is therefore about afterness as such. Inspired by Walter Benjamin's thoughts on translation as afterlife, *Nachleben*, and Derrida's development of this and related topics, Richter examines in his book the meaning of coming and living after, and not least what sort of nearness to what has been and what can be an after carries in itself.

The figure of thought of an after carries, we should add, an uncanny repetition. An after is an after that comes after an after that comes after an after . . . "After" is the figure of a *repetition* that repeats itself. A repetition that repeats not only "something" but the repetition itself defines a lingering note, an echo, a peculiar delaying, peculiar precisely because each "after" delays a previous delaying and creates an odd cluster of several different delayings—an after after an after . . . "Afterness" defines, in fact, the is-being, the movement of the present itself, and not only the delaying of the past or the extending of the now into a future. In this sense, it differs from Gadamer's concept of *Verweilung*, of tarrying, which he understands basically as the remaining of a structure in the present.[19] Afterness is a concept that sheds light on the present's uncanny movement, on how it is near-being[20] insofar as it is-being. Thus, the present is not merely a turning point from the past into the future, but is a movement that continues moving in itself, a co- or multimovement of different delays and echoes. *It is a movement that moves without moving the movement*, a sort of "immobile becoming," to recall an expression by Maurice Blanchot. The perception of the present as either a punctual now that never is or as eternity, in the sense of what always is (*nunc stans*), is blind and deaf to the movement of the is-being, a movement without where-from and where-to, the movement of a *meanwhile*.

The point to be developed on the basis of these reflections on afterness as near-being to the is-being is how, in the experience of exile and exilic memory, the experience of afterness reveals, like a photo negative, the present as a movement in itself, as a moving between and meanwhile, a movement marked by a repetition that can be compared to the echo and lingering of several lingering delays. In question is the temporality of the present as meanwhile and how the meanwhile exposes a nearness to the happening of existence, to the happening *while* it happens. In this approach, exile is not in first place the strong presence of an interrupted past nor the shadow of a threatened future. Exile is rather an immense struggle for presencing, an utmost *present tension*. Afterness shows how the present's movement is a tensioned and disquiet unmoving movement of the between and of the meanwhile. This is what memory bears witness to in and as exile. Because discussions on exile stress on the relation between exile and memory and moreover on the assumption that memory is the capacity to bring the past to the present and the present to the past, it is opportune to indicate how the movements of memory reveal even more strongly albeit in a diffuse way the disquiet of the is-being.

The Exile of Memory

Memory is understood primarily as bound up with and bound to the past. Memory shows how the past can be present. Different theories about memory, from the ancient Greeks to today, have paid attention to how memory not only forms a bond with what has been, and with what is missing, but how it is also a way of making-present. These two aspects are central for both Platonic and Aristotelian accounts on the phenomenon of memory, even if each one gives priority to one of them. While Aristotle stresses memory as "enrapture to the past" (*he de mnéme tou genoménou*),[21] Plato, and later Augustine, insists that memory is above all the present representation of an absent thing.[22] Thus, memory has long been considered a double movement: memory enraptures with the past by preserving the trace or absence of the past, so that it can be called to memory and thereby become a presence in the present. Memory is partly a passive storage and an active de-estrangement. Plato and Aristotle have confirmed this duplicity of memory and have used, with different emphasis, two different concepts to describe it: *mneme* and *anamnesis*, *memoria* et *reminiscentia*, memory/ remembrance and recollection, a difference that has followed the long history of philosophy, and one that received in the German romantic distinction between *Gedächtnis* and *Erinnerung* a decisive significance. In this long history of philosophical reflections about the duplicity of memory, a central issue is how active memory, what Plato and Aristotle calls *anamnesis*, has defined the activity of thinking as afterthought and reflection and even as a force of connection and association between different ideas. The philosophy of Derrida can be considered a radicalization of this traditional identification of thinking and recollection [*anamnesis*]. He stated that if at the heart of his thought of deconstruction, that is, of the activity of thinking as deconstruction, there is a constant experience of loss, there is however one sole loss that he would never be able to bear: namely, the loss of memory. Loss of memory has not only the meaning of amnesia but primarily that of an erasure of traces.[23] As much as in Plato, for Derrida, philosophy is understood as *mnemophilia*, and maybe even more as *mnemosophia*. If, for Derrida, philosophy begins in a transcendental loss, then there is, according to him, a loss that cannot be lost, namely, the trace of memory. He admits even in the same passage that his "wish is not to produce a philosophical work or a work of art but it is only to preserve memory."[24] Derrida's philosophy is the one of a memory of memory, of a mnemonic imperative: do not forget memory.

Since ancient times, memory, both in the sense of passive storage and active becoming-presence, is connected to image. Memory is an image of the lost, of what has been, that both preserves it and makes

it present. Two metaphors or images have been connected since then
to memory: on the one hand, memory as a space, a chest, a box,
a cabinet, a "store-house" (Locke), an archive, a museum, and, on
the other, memory as a seal, a copy or transcript. Memory has been
described in terms of writing down and noting. The work of memory
is as imitative or mimetic as writing itself. However, since memory
not only is a preserving and registering activity but also a creation of
images and therefore a rendering present of what is absent, memory
is not only similar to writing but also to reading. Memory is at the
same time similar to writing and to reading. It is both reproductively
productive and productively reproductive. It is in memory that the
mimetic dynamics of an image most clearly appears. It is also only
because memory is imaginative, that is, capable to create images that
it is possible to speak of cultural and collective memory. Cultural and
collective memories exist only insofar as a past event that occurred in
the life of a group or community is seen and presented from a now-po-
sition [Halbwachs].[25] It is because those memories can be *made present*
that they always implicate a construction and an interpretation and
thereby can be an object of ideological and political manipulations.

    These two sides of the work of memory: preserving and creating,
the keeping of the trace of the past and the making present of what
has been—is however always said in present time: they happen insofar
as memory *remembers* the past. Even if it is possible to remember that
we have remembered, past remembrance can only be remembered in
the present tense, in tension with the present, something that Husserl
described very accurately.[26] Since ancient times and especially in modern discussions about memory it has been taken as given that memory
differs substantially from perception, which has been considered the
sense for the present *par excellence*. Aristotle claimed that memory
does not perceive since its object is absent. In the above named
dissertation about memory and recollection, Aristotle insisted that:
"memory refers to the past since nobody can remember the present
while it is present (*tó de paròn hote párestin*)."[27] According to him, no
one can remember white while you look at it or remember theorizing
while you theorize. The first can only be perceived and the second can
only become object of knowledge.[28] Memory differs from perception
as much as it does from expectation, whose object also is absent but
in an entirely different way. The past has a present absence insofar as
it leaves traces. The future is however an absence without traces, and

can therefore neither be remembered nor perceived but only foretold. At the same time memory shares with expectation the ability to create images, to make the absent present, being, in this sense, a dimension of illusion or of imagination. From Plato to Husserl, a recurrent trope in different philosophies of memory has been the one of similarities and differences between memory and fantasy, between the ability to remember and the "force of imagination"[29] (Sallis).

Even if several philosophers have tried to show differences between images of memory and visions of future as the difference between an understanding of images as signs and traces and of images as imagined objects (a distinction developed by Diderot),[30] they have admitted a closer relation between memory and imagination than that between memory and perception. However, it is first with the increased awareness of the modern condition as a condition of losses, of a transcendental "no-longer," of a time that has disrupted the continuity of history, that the role of memory in perception and thereby in now and present awareness has reached thematic emphasis. Romanticism recognizes in memory a transcendental dimension where existence is comprehended as a fragment of memory of the absolute, that is, of the ancestrality and "faraway-ness" of life, as a memory, a "*Fernsichtigkeit*," of "farsightedness," to recall an expression by Novalis.[31] Presence, now-being or near-being is from this romantic viewpoint understood as recollection of the eternal life of the absolute, as its trace of memory or fragment. Memory appears thereby as memory of memory itself, a memory of not only *what* has disappeared but of disappearance as such, a vision that Hobbes, much earlier, tended to share when affirming that memory is the expressivity and significance of the "fading" of the senses."[32] It is though only when the now and the present, when presence itself is experienced as catastrophe, that is, as an interruption of life where a "constellation of dangers," as Benjamin expressed, emerges as vintage point for experience, that the figure of thought of "a memory of the present" [*une mémoire du present*], to borrow the words by Baudelaire, becomes possible. The experience of exile can be formulated as a present and a presence that no longer can be perceived or imagined but must be continuously remembered. The experience of exile denies the Aristotelian conviction that no one can remember the present while it is present (*tó de paròn hote párestin*) while it is being. *Exile is indeed memory of the presencing of the present rather than the presence of memory.*

That a present is rather remembered than perceived, this is a central observation underlining Proust's and Freud's mediations on memory. According to them, but in different ways, memory both makes possible and eludes perception, for memory is responsible for both opening and blocking the senses, for example, because of a trauma. Memory is understood either as what reveals the present or what covers it over. Both Proust and Freud present the memory of the past and the perception of the present as a tension between background and figure, where the figure of the present is moved to the background and memory becomes figure and/or the reverse. In exile, however, the perception of the present is already a memory and not only the inescapable connection between consciousness and the subconscious, between pursued and suppressed memories, or between "voluntary" or "involuntary" ones. That perception in exile necessarily is already memory lies in that, in exile, the is-being is neither here nor there, but rather the nearness to its inescapable faraway. It shows that in some way, memory is always the trace of another, indeed of the other and of the others. As Derrida insisted, memory is always a "memory of the other,"[33] because memory is the experience of a certain "sorrow," *deuil*, of the others' absence. This sorrow, *deuil*, Derrida understands as care, as caretaking of losses and separations, of afterness as such. In this regard, what is seen, heard, and touched—that is, the present—is always already the trace of another, showing in itself different traces. Derrida's insistence that Western metaphysics is a metaphysics of presence, aims to stress how the Western philosophical tradition has been unable to see that the present is never present but only the trace of an afterlife. This is the basis for his view of memory as what does not remember firsthand what has been present but rather what can never become present. According to Derrida, the condition of possibility for memory is that memory is continuously remembering what can never be present, the loss and disappearance that constitutes every presence. For Derrida, under a Benjaminian influence, the present is a trace and thereby the memory of something other, of the disappeared, the quieted down, the oppressed, and therefore of the forgotten. In his thoughts, we can recognize echoes of St. Augustine's discussions about how memory remembers oblivion, and thereby how "memory retains forgetfulness" so that memory does not forget to remember.[34] Derrida's thoughts about memory, about thinking as remembrance, depart from some postulates: that presence is what remembers because it is a trace and not something that exists; that traces recall to mind; that memory is always memory of the other and if the "I" that remembers is always remembered in "my" memories it is because I myself am nothing else than the traces of it and of the other. These postulates require however another, let's say, transcendental postulate, namely, that memory remembers what can never be present, that is, presence itself. Presence is therefore re-defined as différance and not as a difference from what has been. If Derrida tried to eradicate from thinking the Western obsession with consolidating presence into a now-point and leaving behind the vocabulary of present and presence it was indeed for the sake of remembering that presence is nothing but difference, nothing but difference differing from itself.

Derrida's understanding of memory as deconstruction and deconstruction as memory of what has never been and can never be present—that is, of presence as différance—can be read as an attempt to show how presence is always after itself. This describes a decisive aspect in the experience of exile. Thus, in exile the present and its presences are always after itself/themselves. Different theories, philosophies, phenomenologies, and even deconstructions of memory have continuously disregarded *how* memory remembers. Much has been discussed about *what* memory remembers and *who* remembers,[35] about the fact that memory constructs identity, both individual and collective, about how politics of identity manipulates, uses and abuses memory. Much has been investigated about how everything can be remembered through technology and how the gigantic amount of memories produces memories by itself. Research on Alzheimer's disease and other kinds of pathology deriving from loss of memory increases day by day. Nonetheless, a view on the embodied temporality of the event of memory is still lacking insofar as a comprehension of the *meanwhile*, of the *enduring* of memory is still to be unfolded. An understanding of how memory remembers in the condition of exile can make a contribution to such a reflection.

Few writers have focused so clearly on the *how* of memory, on memory while remembering, on the event of memory in the experience of exile as Vladimir Nabokov. To Proust, he adds the exilic perspective. Nabokov is well known as one of the greatest writers of exile in the twentieth century, writing not only on and from exile but also in several languages, which means indeed writing between languages. Between languages is the language of exile. As he wrote in his autobiography, his literature is the literature of memory speaking. *Speak, Memory* is how he entitled his literary autobiography.[36] In exile, memory is what speaks and not what someone, the author, writes about. The imperative of exile is precisely that which demands the acceptance of living under the speaking of memory.

Nabokov describes in his work how, in exile, the present appears as nearness to being in the very happening of memory. Thus, in exile, everything that "is" appears as an "it has become" with such an intensity that what has been can no longer be distinguished from an "it has become something else" and from the expectation that it was and could have been otherwise. Moreover, the expectation of what can be and could be goes hand in hand with the fact that what has been was abandoned and rendered strange. Everything that was, that is, and that can be appears as such in the light of the separation and leaving behind, appearing thereby as an on-the-way toward uncertainty, that is, as movement, a disquiet, worried and moving movement.

Everything appears with the without and without a with. This is due not only to what Edward Saïd once wrote, namely, to the fact that "in exile, habits of life, expression or activity in the new environment inevitably occur against the memory of these things in another environment,"[37] which happens to everyone and belongs to every kind of memory. Nabokov realized more precisely what happens with exilic memory. Memory in "exile" is not simply a memory that sees two things at the same time—what appears now and what has been seen before, a memory that has two words for each thing, a word from the own language and a foreign and sought word. He realized that the memory in exile is a "shimmering go-between," a movement of several lingering notes, echoes, delayings, escapings, and afterimages, indeed, of what can only be seen while disappearing like a comet and an eclipse. Memory in exile shows how memory is perception of a meanwhile, of the meanwhile of a presencing and abiding. That is why Nabokov considered that in exile memory neither retrieves nor creates images. Indeed, he will insist that it is not really about images, but about what he called "photisms,"[38] from the Greek *phos*, light, the shimmering of light forms, rendering visible the on-the-way-toward and from in everything that is and is not. "At times, however, my photisms take on a rather soothing *flou* quality, and then I see—projected, as it were, upon the inside of the eyelid—gray figures walking between bee-hives or small black parrots gradually vanishing among mountain snows, or a move remoteness melting beyond moving masts."[39] Memory in exile, exilic memory, is not only double as all images are, or a mere simultaneity of different fixations in time. It is rather a multidirectional movement that does not inhabit two worlds but that habits the "shimmer go-between" or between-movement of several different betweens. What memory remembers in exile is afterness itself, the movement of existing between worlds, languages, images, sensibilities, indeed, the movement of between-existence. Nabokov considered that in exile one must "go through comets and eclipses"[40] since everything that is seen, heard, said, thought, and touched is heard, said, thought and touched while disappearing into the appearing of something else. With the concept of "photisms," Nabokov aims to indicate that, in exile, what is remembered is the coming and going of images rather than images themselves. Remembered in exile—conscious or subconsciously—is the leaving behind, the flight, the meanwhile and enduring of afterness itself. In this direction, it is possible to understand why Nabokov's exilic memories have to be kinesthetic and why, among the different life-forms, it is for him the butterfly (he was a great connoisseur and collector of butterflies) that most clearly embodies a transformation while being transformed, the is-transforming in its continuous form, the presencing rather than the present.

Memory in exile bears witness to how exile is the experience of an after that comes after the after that comes after, and as such, the experience in which the present appears as nearness to the is-being, as almost being, a nearness to multidirectional movements. Exile is perhaps nothing but "approaching" (agchibasien),[41] recalling a lonely word attributed to Heraclitus. It shows that it is within several layers of back-and-forth-movements, within the diverse and clusterlike between of multidirectional movements, that the present and its presences appear as nearness to the is-being. The question of exile today, the question about the refugees in a world running faster and faster toward violent forms of politics of segregation and exclusion, should be discussed on the basis of the way the present and its presences are experienced as "shimmering go-between," showing how existence itself is an on-the-way-toward the is-existing, without return and without arrival.

These introductory reflections on the times of excess and exile and on the features of memory in exile aimed to indicate the need to think otherwise some basic presuppositions that orient most thoughts, theories and literature of exile today. This book aims to challenge basic theoretical views on exile without any attempt to account for concrete experiences of it. More specifically, most of these theoretical views presuppose that to exist in exile is to exist after a cut and interruption and that exilic existence is fundamentally the memory of this cut piercing every dimension of life and death. The above discussions attempt to motivate the need to follow another path of thought in relation to existence in exile. This other path is the one of speaking and thinking about existence in exile departing from the liminal experience that exile is. From within exile, what prevails is neither the memory of a past overexposed in the present nor the expectation of a future fading away in the present. Exile rather exposes existence to the is-existing of existence, revealing the temporality of exile as gerundive. In the next chapters, a tentative effort will be made to develop a thought of the gerundive temporality exposed in the experience of exile following some traces, drafts, and embodiments of it in conversation with Heidegger, Blanchot, and Clarice.

Chapter 2

The Ecstasy of Time

(In Conversation with Heidegger)

Time is not a thing, thus nothing which is, and yet it remains constant in its passing away without being something temporal like the beings in time.

—Martin Heidegger

The Ecstasy of Time in *Being and Time*

It is far from clear why a reading of Heidegger could contribute to a thought of "time in exile" and more specifically to a thought of the gerundive temporality of exile. "Exile" is not a word found in Heidegger's work. The closest to it are words such as "homelessness" [*Heimatlosigkeit*],[1] "up-rootedness" [*Entwurzelung*],[2] "groundlessness" [*Bodenlosigkeit*],[3] "the uncanniness of unfamiliarity" [*Unheimlichkeit*],[4] "worldlessness" [*Weltlosigkeit*],[5] "never dwelling anywhere" [*Aufenthaltlosigkeit*][6] and "unworlding" [*Entweltlichung*].[7] Indeed, Heidegger is the philosopher who stayed proudly in the province, not only in the 1930s but all his life, affirming "the inner relationship of his work and the rootedness in the Alemannian-Swabian soil."[8] Moreover, he is a philosopher who never abandoned the need to find a thoughtful way to dwell in the up-rootedness of the epoch of Being he calls *Gestell*, translated as "positionality" or "enframing."[9] If not a philosopher of exile, however, Heidegger is, and very much so, a philosopher of the excess, and indeed *the* philosopher of the excess of philosophy. Beyond being a philosopher of the excess of philosophy, Heidegger is also a philosopher of the ecstasy of time and intimacy. Exile is not a Heideggerian figure of thought, although his thoughts of excess and ecstasy have inspired some works on "exilic existence" and "existence in exile."[10] Most attempts to think exile with or even against Heidegger focus on "ecstatic temporality," which Heidegger defined as being "out of itself in and for itself."[11] And it is on the basis of this temporal meaning of ecstasy and of the ecstatic meaning of temporality that Heidegger develops his thoughts on homelessness, unfamiliarity, groundlessness, and worldlessness, above all in *Being and Time* but also later on, though each of these thoughts undergoes significant modifications. If a thought of exile and existence in exile can be found between the lines of Heidegger's thought, it is the thought of a temporal dimension of exile rather than of its spatial determinations. Following Heidegger's thoughts on ecstatic temporality, one finds, however, solid elements for defining exiled existence from within. In this chapter, I intend to show that Heidegger's thoughts on "ecstatic temporality" present an oscillation in which he comes close to a thought of gerundive time, thus contributing to a thought of exiled experience from within. This oscillation is already present in *Being and Time* but becomes stronger and clearer in the writings after the war, especially in the lecture *Time and Being* from 1962. At this point of his itinerary of thought, Heidegger accentuates less the "ecstasy" of being and instead attempts to grasp the presencing of being, touching the nearness to what I am calling a gerundive sense of time and being.

In *Being and Time*, "ecstatic temporality," the trance or ecstasy of an "out of itself in and for itself," does not define a state or a contingent possibility of human life but human existence as the comprehension of Being. As such, "ecstatic temporality" defines human existence as exilic condition. Humans do not exist as things do, in the sense that things are something and are thereby self-sustaining. Indeed, it should not be said that humans exist, if existence is meant in the sense of being as thing, as something, as what has a substantive and substantial meaning. Quite differently, the only thing that can be said is that humans exist not as a thing among other things but as a certain kind of verb, the verb "to exist," formed by the prefix "ex-," that is, "out of" and *sistere*, the present active infinitive form of the Latin verb *sistō*, the same as the Greek ἵστημι (*hístēmi*), that is, to stand in itself. Heidegger finds in the German verb *da-sein* the direct German correspondent to the Latin verb *ex-sistere*, to exist. Indeed, *dasein* is a verb, and the "ecstatic temporality" that constitutes *Dasein* is the temporality of being verbal rather than a substantial, taking place in time and taking time in place. The huge difficulty in translating the use of the term *Dasein* in *Being and Time* lies in the way *dasein* is a verb which is discussed in substantive terms.[12] This becomes both clear and insoluble above all in the second part of *Being and Time*, which deals with temporality in general and more specifically the "ecstatic temporality" of the verb *dasein*. The difficulties in speaking the language of "ecstatic temporality" are so big that Heidegger has to interrupt the book and leave it unfinished as a "torso." In order to consider the basic aspects of these difficulties, it is important to consider not only that *da* in *dasein* "translates" the *ex* in existence but also the philosophical sense of Heidegger's way of adding prefixes to the verb of all verbs, the verb "being," *sein*. As Levinas illuminatingly observed, what Heidegger does in the hyphenated expressions—*in-sein*, *mit-sein*, *da-sein*, *für-einander-sein* and so on—is to add prefixes to the root of the verb "to be." Developing this Levinasian indication, we should understand these terms rather as new verbs and think them as "to in-be," "to with-be," "to there-be," more *zu dasein* than *dazu-sein*. That might explain why Heidegger, in *Being and Time*, uses the hyphen separating *da* and *sein*, a hyphen that becomes even more emphasized in the writings of the 1930s like *Contributions to Philosophy*. In play here are the difficulties of describing the ecstatic temporality of human existence as the infinitive form of the verb "to exist," of *dasein* as a having to be,[13] an infinitive form situated in the very fact of existing. This is why Heidegger says that "we can come to terms with the question of existence always only through existence,"[14] quoting Joan Stambaugh's translation, which, according to the German original, should rather read, "[W]e can come to terms with the question of existence always only through the to-exist [*das Existieren*] itself." Thus, it is not through existence, that is, through its substantive meaning that the "verb-character" of human condition becomes clear, but through "the to-exist," *das Existieren*, through its infinitive verbal form. *Dasein* means being a to-be, existing as a to-exist. The claim is that *dasein*'s determination, the way the human exists, is an open indetermination or infinitiveness, a to-be and to-exist open for and even demanding continuous determination.

The vocabulary of existence is very nuanced in *Being and Time* and sometimes oscillates between the infinitive form "to-exist" [*zu existieren*], the substantivized form of the present participle "the existing" [*das Existierende*] and sometimes even the gerundive form "existing" [*existierend*]. Heidegger uses these forms without consistency. Existence is to be distinguished from *what* exists, in the current conception of something present at hand (*vorhanden*); rather Heidegger is concerned with *how* existence exists. It names human existence as distinguished from what "exists" insofar as it is nothing determined and defined. It is instead an opening, a being-possible, a "care," and, as such, a being-thrown into the is-being of being and not what is grounded on the Being of beings. Existence exists in-being, *in-Sein*, in the openness of being. In order to make the distinction between existence in the sense of "what" exists (*existentia*, presence-at-hand, *Vorhandenheit*), and existence in the sense of "how" existence exists in exposure to the openness of being, Heidegger speaks sometime of "*existing* Dasein," existing existence. As "existing" (*existierend*), Dasein can never be ascertained as a given matter of fact that appears and disappears "with time" and that in part is already passed.[15] Existing Dasein, which is said in the text both as *Dasein existierend* and *das existierende Dasein*, defines the "existent" [existential] or "ontic" dimension in which the existential apprehension of "existence in-being" is rooted.[16] This existing is the "root" for apprehending existence as existence in being. Derrida, in the seminar he held in the 1960s on *Being and Time*, focuses on the question of being and history and places particular stress on Heidegger's claim that the existential understanding of existence should be "rooted" in the "existent" and ontic dimension of existence—a problematic claim according to Derrida.[17] What is problematic, however, is how "existing Dasein," existing existence, is understood. If this rootedness is conceived of as rooted in *existing*, it has to do with roots in the event of existence, that is neither in space nor in time, neither in a place nor in an hour, being nothing and everything, as "a quality of loss/ Affecting the content/ As Trade had suddenly encroached/ Upon a Sacrament" to describe it with a verse by Emily Dickinson.[18] It points to being rooted in what has no roots. Existence—which has no substance—is the substance of the human.[19] Existence, as Heidegger insists, is the essence of *Dasein*. These definitions are tautologies, and what they say is indeed that existence is the essence of existence, that existence is nothing but existing. Even though Heidegger does not formulate it in these terms, the oscillation between the terms "the existing" (*das Existierende*) and "existing" (*existierend*) points toward such direction of thought. This understanding of existence does not derive from any idea of existence. In section 63 of *Being and Time*, right before the analysis of the temporality of Dasein, Heidegger asks where this evidence comes from. It comes from the evidence, in the sense of what shows itself from itself, of the being ahead of itself of existence, termed *Sorge*, "care." The being of existence (*Dasein*) is being ahead of itself (*Sorge*), and this is nothing but the existing of existence. This is why

> Existing (*Existierend*), Dasein never gets back behind its thrownness so that it could ever expressly release this "that it-is-and-has-to-be" from its *being a self* and lead it into the there. But thrownness does not lie behind it as an event, which actually occurred, something that happened to it and was again separated from Da-sein. Rather, as long as it is, Da-sein is constantly its "that" (*Da*) as care. As this being, delivered over to which it can exist uniquely as the being which it is, its is, *existing*, the ground of its potentiality-of-being. Because it has *not* laid the ground *itself*, it rests in the weight of it, which mood reveals to it as a burden.[20]

We must stress here the "that it-is-and-has-to-be," thrownness, on the one hand, cannot be released (*entlassen*) from its being a self (*Selbstsein*), but on the other hand, it cannot assume the self as the ground of existence. *Existing* is the ground of existence. *Existing* is "being the ground." It is a groundless ground for it has neither bottom nor end, neither breadth nor depth, and it overwhelms the whole of existence. It overwhelms the whole insofar as it is *ahead* of itself, like a shadow one wants to grasp. It is the *Da*, Heidegger says, the "*that* it is," and it is from out of this "that it is," from out of *existing*, that it is a *way* of existence and not *what* exists. Existing, that is, nothing but this "that it is" (*Da*), is a groundless ground that can be grasped neither as a topology nor as a chronology; it has never begun and it never ends; it does not come from anywhere and does not go anywhere. It is a coming and going without provenance or destiny. "It does not have an end where it simply stops, but it exists finitely."[21] Finite existence means existence in being. It is being. *Existing*—the groundless ground of existence—is neither spatial nor temporal, for it is gone while coming, it comes while going away. It says itself; it exposes itself in a kind of present, in a present participle, in a present ahead of itself. Heidegger touches here the gerundive sense of time. Considering that the Latin word *Praesens*, build from *prae-sens*, means "being ahead of itself," as shown by Emile Benveniste and Henri Maldiney,[22] we have here to do with a mode rather than a *tempus*. Existence in be*ing*, exist*ing*, that is, being ahead of itself, is *ahead* [*vorlaufend*] insofar as it passes by without passing away, exposing itself as a discontinuous continuity of while-ness and between-ness. It would, however, be misleading, says Heidegger, to "orient" the discussion towards the "between," and we could add towards the "while" too, if, in doing so, they would express the "result of the *convenientia* of two objectively present things."[23] What more clearly grasps the relation, we could say, is the hyphen, which means the "under one," that brings together the "having been-being-coming-to" ["*gewesende-gegenwärtigende-Zukunft*"],[24] a formula saturated with the present participle. Heidegger understands this in-one-another of *tensions* (rather than "states") as the meaning of the Greek ek-stasis, ec-stasis, used philosophically by Aristotle in his discussions about the nature of time and change. The ec-static "out of itself in and for itself" defines the "ex" of existence and indicates the sense in which existence is "ec-static temporality."[25] The in-one-another of tensions, the discontinuous continuity, of existing, defined formally as "having been-being-coming-to" is understood here from out of the "ex" of existence, from its ecstasy or excess. The expressions "having been-being-coming-to" and "out of itself in and for itself" aim to express the impossibility of separating the having-been, the being, and the coming-to-be. "The existing" cannot be grasped topologically or chronologically for it is an enigmatic whole, which in its turn can neither be separated into parts nor apprehended as a whole per se, that is, as a whole apart from the parts.

Existing, the *is-being in-being* is a whole that indeed cannot be conceived of as a whole for it has always-already overwhelmed existence. In this sense, it does not mean a "whole" but the "entirety" of existing, of being. The question of how to conceive the existing of existence, the is-being of being is the question of how to grasp what has always already grasped existence. What is to be grasped has already grasped us. "Ec-static temporality" is an attempt to name this strange noncircular circle. It is a circle because it has no way in and no way out, and yet it is noncircular because it is itself without referring back to itself. Indeed, the whole of *existing* means the world, not the world outside or inside, not the world around or beyond, but the world as the event of existing existence. And if here "spatiality is attributed to it in some way, this is possible only on the basis of the in-being."[26] What then is the character of this spatiality, in some way "attributed" to the in-being? The answer given by Heidegger in *Being and Time* is: "[T]he character of de-distancing and directionality."[27] At the basis of the schema of orientation which, since the Greeks, has been oriented by the topological and chronological meaning of existence—as from-to and before-after—we find the "animal with eyes," the "eye-point" [*Punktauge*]."[28] The schema of orientation is based on "points of views," where the "eye" locates the "I" as absolute "here," as the middle point from which every "there" can be measured. Kant will develop this eye-I-centered schema of orientation, showing that orientation in space depends on a "subjective feeling" from which a right and left can be distinguished, from which oriented space can be constituted, and from which the idealized geometrical and logical homogenous space can be deduced.[29] Oriented space is the space of "what" exists, which presupposes a self-point of reference.

As a whole, however, *existing*—that is, existence in being, the event *of* the world rather than events *in* the world—is de-distancing and directionality itself. Existing is therefore neither here nor there, neither before nor after, but instead tends to nearness, and, as such, de-distancing. "An essential tendency toward nearness lies in *Dasein*,"[30] being ontically the closest but, "precisely because of that," ontologically the most distant.[31] Existing, in other words, is neither a direction nor something directed from or to, but rather a tendency, a tension, a "tense," described by Heidegger as directionality [*Aus-richtung*]. In this sense, it is no "self-point" [*Selbstpunkt*][32] because it disrupts the topological and chronological categories of *from-to* and *before-after*. In the sense of having been-being-coming-to, of "out of itself in and for itself," Heidegger insists that existing has no "self-point." To the question of how to grasp this whole that is an overwhelming "happening" [*Geschehen*], there is no self-point from which a sight can be accomplished. The sight of existing can only be the sight *of* existing—for there is no inside or outside from where a sight could be described as near or distant. This "point of view" of existing is neither a point nor a "self," much less a "view." It is the sight *of* the world, the sight "from within" the event of existence existing, the sight from within the intimacy of being. Heidegger calls this sight "transparency" [*Durchsichtigkeit*][33] and opposes it to "self-knowledge," leaving here a draft for a "destruction" of the "center" of philosophical thought, namely, self-knowledge, that is, the knowledge oriented by the idea of the self, *to auto*. Instead of a self as the most primordial point of reference or orientation, a "self-point," existing exposes a "throughout," *durch*. Here there is no self, *auto*, but a "through-out (*dia*) sight," [*Durch-sicht*] that "corresponds to the clearedness [*Gelichtetheit*] characterizing the disclosedness of the there [*Da*][34]—of "that it is existing." The seeing of this sight is neither bodily nor spiritual; it has no eyes, neither of the body nor of the soul.[35] Heidegger will call it under-standing, *Verstehen*, in the sense of standing in this "throughout," in this tensional between, a sense that perhaps not even he himself could grasp entirely, but that is better captured in Jacques Derrida's and Jean-Luc Nancy's discussions about "touching."[36] Transparency, *Durchsichtigkeit*, defines the sight of existing, the sight of the tensional between and its basic temporal mode of "meanwhile."

As a result of the previous description, it can be said that in *Being and Time*, existence in-being is neither oriented not disoriented but rather through-oriented, when this "through" is understood as the tension of between. It is through-oriented (we could also say per-oriented) not in the sense of being oriented towards an outside or an inside but in the sense of being out of itself in and for itself—a tendency to proximity, a de-distancing, indeed an approaching. In the attempt to understand from "within"—to interpret existence in-being, *existing* existence, the ungraspable and unnamable having been-being-coming to be, Heidegger formulates a phenomenology of the "ex," of exposure, excess, and ex-orientation, of the ecstasy of existence. Putting the accent and emphasis on the "ex," the existing as such—in its verbal mode—comes to oblivion. In the emphasis on ecstasy and excess of being and existing, the existing as such is dismissed and subsumed to the thought of the infinitive "to exist," which implies a thought of the "to-come" and "to-overcome," Heidegger's ecstatic temporality is infinitive temporality, assuming basically the is-being rather as an infinitiveness in being rather than the gerundive mode of is-being.

The Ecstasy of Overcoming after *Being and Time*

Dasein is never in itself, but always out of itself. It is however out of itself in a particular sense, namely as a coming to itself. But how to define the "self" of this out-of-its-self that *dasein* always is, which can no longer be grasped as any "self-point"? Here lies a central difficulty in Heidegger's understanding of *the self*, and indeed its blind spot. On the one hand, *dasein* is ecstasy, is ek-sistence, to say it with the hyphen that Heidegger uses even more frequently and decisively after *Being and Time*. *Dasein is* continuously out of itself, implying that *dasein is* not itself in itself. This means, on the one hand, that *Dasein* is *not* a self and, on the other, that it *is* a not-itself. This means that traditional concepts of the self, both ancient and modern, that take the self to be either substance or subject, essence or process, do not fit to the way humans exist as a to-exist, the mode of being a to-be. Being, moreover, a not-itself, *dasein* permanently misunderstands itself if it conceives of itself as a substantial, essential, or processual "self." This continuous misunderstanding, which indeed corresponds to philosophical conceptions of the self, confuses the "self" with *Man-selbst*, with "oneself," with "the they." Heidegger's analyses in *Being and Time* of *das Man*, of the "they," are "concrete" descriptions of the logic operating in traditional conceptions of the self as substantive identity and subjectivity and thus not merely discussions about decadent, improper, and inauthentic ways of existence. They are moreover an "existential" account of the subtle difference between being not oneself and being a non-oneself. If dasein *is* in itself out of itself, if it *is* ecstatic, and if "out of itself" means both not a self and a not-itself, that is, if inauthenticity belongs to authenticity, then how are we to understand the call for "coming to itself," which Heidegger terms the "call of consciousness?" How to understand the imperative: "be" or "exist" implied in the infinitive modes "to be," "to exist"? To which "itself" shall *Dasein* come to? The blind spot of Heidegger's concept of the "self" is that the "coming to itself" is for Heidegger precisely the authentic self. *"Coming to itself" is the "self" to which dasein is appealed to come to.* In the vocabulary of *Being and Time*, this is discussed in terms of care, *Sorge. Care* is the "formal indication" of Heidegger's understanding of the "self" beyond ancient and modern conceptions of selfhood and subjectivity, including romantic-idealistic conceptions of becoming and phenomenological accounts on intentionality. That is why Heidegger insists that *care* should be understood neither as intentionality nor as "drive," "wishing," or "willing."[37] *Care*, let us say, "the Heideggerian self," has no positionality, being rather a structure of adverbs and pre-positions. In the vocabulary of *Being and Time*, *care* is defined as "being *ahead* of itself *already* being in (a world) as being-*together*-with (beings encountered within the world)."[38] "Ahead," "already," "together," "with"—each one of these adverbs expresses an ecstasy, an out of itself in and for itself, and thereby a "steadiness that has been stretched" [*erstreckte Ständigkeit*][39] or "between." In *Being and Time*, it is said that "as care, dasein *is* the between,"[40] a between that *is* itself a dynamic whole, a stretching-out and not an interval between points. In these terms, care aims to describe the sense in which the coming to itself is the self to which one is to come. Heidegger also calls this "anticipatory resoluteness," defined as "being toward one's ownmost, eminent potentiality-of-being."[41] But once again, whose ownmost, eminent potentiality-of-being"? The answer is indeed the potentiality-of-being potentiality-of-being, that is, of being in such a way that potentiality can endure *as potentiality* and not be destroyed or pacified by any actualization. Or put otherwise, to be toward being-toward. These reduplications and tautologies aim to express the difficult figure of thought of the self that Heidegger proposes, the one of becoming the potentiality-of-being one always is already. Following Levinas again, particularly what he said in a conversation with Jean Wahl in 1947, it is the awareness of itself as being-toward-death that enables *dasein* to be toward being-toward, toward being the potentiality of being and not the actualization of some potential. For Levinas, this thought of Heidegger is the fundamental meaning of Heidegger's being-toward-death. This expression gives the true meaning of Heidegger's thought of possibility as primary, according to Levinas, thus breaking with the Aristotelian frame of thinking potentiality on the basis of actuality, in which potentiality is that which actuality has to overcome in order to become actual.[42] However, the blind spot of Heidegger's thought lies in the fact that "coming to itself," "being-toward-itself" or "one's own most and eminent potentiality-of-being" each becomes itself a substantive form.

The contradiction in terms of a self that is a coming to itself, of being toward a being-toward, of being the potentiality-of-being appears more clearly in the dark thoughts about the "people," about "the Germans to come" that Heidegger formulates in the 30s. The

Germans, or the people of the "other beginning" as Heidegger will speak about in the *Contributions to Philosophy* and the *Black Notebooks*, are not those who exist, but instead the only ones that could exist as the potentiality-of-being, those who could be *the* coming-to-be. The difficulty lies in this substantive form of coming to be *the* coming-to-be, in which the ecstasy of being itself becomes ecstatic, indeed hyperbolically meaningful, where the self that is not a self and a not-itself is no longer itself and appears as an excess of subjectivity, of selfhood. In these dark thoughts, we find the figure of the ecstasy of ecstasy, of the excess of excess when the ecstatic becomes hyperbolically static. Heidegger is undoubtedly an excessive thinker, a thinker of the excess.

This excessive meaningfulness of the self that is not only out of itself but is *the out* of itself, finds its basis in the privilege of the to-come over of the present. This privilege, in turn, is based on the originary and primordial status of the "to come toward," which is the ecstatic dimension of what is normally called the future. The privilege of the "to come toward" is related to the entanglement of "to-come" and "toward," that is, of "to-come" and a direction, more precisely defined, a destiny. Even if destiny is understood in *Being and Time* as distinct from destination and from a sense of movement defined as teleological, there is a view or a guess about what to come, about a coming-over. Or to say it in another way: the basis lies in Heidegger's understanding of the to-come as *overcoming*. It is the pathos of revolution, of an overcoming that pushes the thought of existence and of being toward the ecstasy of excess.[43] Especially problematic here is the understanding of the verbal meaning of the verbs "to exist" and "to be" as coming toward, as overcoming. The problem is that the to-come and its appeal to overcome steps over the is-existing of existence, the is-being of being; it steps over the gerundive structure of existence more than the present or what is present. At issue here is the inherent ambiguity of the verbal aspect of the "to-come" and thereby also of the verb "to overcome." Thus, saying in present tense, "I come," we are also actually saying a future; in the "to come" the future is said in the present tense.

What is to be overcome is the forgetfulness of being—this is, as is well known, a central motif of Heidegger's philosophical narrative. "Forgetfulness of being": but in what sense? Forgetfulness of being as being, when being is taken as what is being, as "presence," as ground and reason of the beings. Being as being: this means, being as a verb,

and as such in a certain sense as "time." For Heidegger, this forgetfulness is not only constitutive of the human way of existing, of *da-sein*,
insofar as it is the way of being out of itself in itself. For him, this
forgetfulness has also grounded a civilizational way of existing, the
Western, occidental way, which is, for him, a civilization grounded on
philosophy, on the question about the being of beings. The question
of being is a human question; thus, to be human means to be qua
comprehending of being, and in such a way that in all ways of being,
the human is already responding to this question. But the question of
being is not only implicit in all human imprints on the life of life: the
question of being became itself a question, in a certain moment, in a
certain language: it inaugurated a civilization. The question of being
became itself a question, however, in a certain mode: it was asked as
a question about the being of *all* beings. In the opening pages of *Being
and Time*, Heidegger introduces a central element in his narrative about
the beginning of philosophy. He says that philosophy begins with a loss;
thus, in asking the question about the being of the beings, it forgot
what enabled this question, namely, the encounter with the enigmatic
fact that being is, indeed, *that* being is being. The fundamental trope
of Heidegger's thought—the question about the meaning of being—is,
indeed, the question about the oblivion of being when being is assumed
as *what* is being, as the substantive meaning of something present at
hand or simply given—as *Vorhandenheit*. In *Being and Time*, an attempt
is made to *destruct* the meaning of being as simply given and present at
hand, as *Vorhandenheit*, by means of an attempt to bring this meaning
back to its verbal sense. The attempts are made through unusual uses
of substantive forms in verbal forms, through the creation of a series
of new verbs when, as we saw, new prefixes are added to the root
verb "to be," and so on. The attempt is to build a new "grammar," as
Heidegger suggested explicitly,[44] a new language to say "being" verbally
rather than substantially. But this is only possible insofar as reminders
of the oblivion of being remain legible in all the inauthentic, vulgar,
universal, and formal thoughts of being that ground traditional ontology and metaphysics. Heidegger remains here quite Kantian, above
all when Kant discusses how reason generates metaphysical "illusion"
(*Schein*), not because of rational mistakes but because of mistakes owing
to the very nature of reason. What Heidegger calls the perspective of
"the they" or "vulgar" and "inauthentic" interpretations is the basis of
complex metaphysical statements. If metaphysics, that is, philosophy,

is the question of being that forgets being, it is indeed a forgetfulness that reminds us of what it forgets precisely in forgetting.

The ontological narrative that Heidegger presents departs from a struggle with the times, with modern times. Unlike Husserl's critique of modern objectification and subjectivism, of the naturalization of consciousness through representation, that for him has forgotten the Greek meaning of philosophy and becomes thereby alienated from the life-world, Heidegger sees the problem in philosophy itself. Heidegger's thought is obsessed with the need to overcome philosophy from within, insofar as he sees the problem in the foundation of philosophy as a universal search for foundation that enabled the foundation of a universe grounded on the search for becoming more and more universal. The crisis of modernity is the crisis inaugurated by philosophy as metaphysics, that for Heidegger means the oblivion of being in the determination of being as *what* is being and not only the separation between the sensible and the intelligible or between the visible and the invisible. With Heidegger, the question of metaphysics is no longer framed solely in terms of the dichotomy between sensible and intelligible but as the "ontological difference" between being (as verb) and beings (as substantive meanings). It is the reduction of the verb character of being to substantive forms, to thinghood, to objectivity, which in a Marxist vocabulary could also be termed as the reduction of being to reification and "commodification." At stake in the problematic nature of philosophy is indeed the oblivion of the "ontological difference" between being and the beings as, to a certain extent, inescapable. The thought of overcoming metaphysics is presented in *Being and Time* in terms of "destruction" of traditional ontology, in terms of the need to show how the loss of the insight into the verbal meaning of being was performed in philosophical thought, deconstructing its layers of construction, not for the sake of undoing what has been done, which is impossible, but of investigating another way of thinking that can emerge as a photographic negative from this scrapping of layers of philosophical tradition. This, however, is only possible if a "turn" (*Kehre*) in *Dasein* takes place, if *Dasein* engages with having to exist authentically in its own existing, or, in other words, having to be the potentiality-of-being; thus, what remains to be found anew is a thinking of the intertwining of life and existence. In question for Heidegger is how to think in a way that is attuned with the meaning of being as being.

What grows, however, in Heidegger's thought after *Being and Time* is the insight into how the forgetfulness of being also reminds of what is forgotten, namely, that being *is*, the meaning of this simple but enigmatic verb. Indeed, what grows and becomes clearer for him after *Being and Time* is how the forgetfulness of being belongs to being itself, is *of* being. This means that the fact that being is not *what* is being but that being is being (and not *a* being) can only show itself while *withdrawing* into the beings. To appear in withdrawal itself is what Heidegger reads in the Greek word *aletheia*, which means truth, *Wahrheit*. That is why, rather than a question about the meaning of being, Heidegger has to rephrase it as the question about the truth of being. Not because there is or should be a truth of being that has been forgotten or that should be grounded anew, but in the sense that being *is* truth, that is, that being as being shows itself while withdrawing into beings, in *what* is being. If being as verb names a happening, an event, a *Geschehen*, every attempt to say being as being, to say being is, or being is being makes the mistakes that Nietzsche recognized in attempts to say the happening in a grammar, the philosophical-metaphysical grammar, that can only speak in terms of a subject, of a noun following the logic of predicative sentences. Nietzsche gives as an example the occurrence of lightning. He claims that in saying "the lightning brightens," two mistakes are made: the first, the one of taking the happening for something being, and the second, the one of taking the happening for an effect.[45] The lightning—the subject—is already the predicate, and the predicate—the brightening—is already the lightning. The same can be said in relation to being; thus, to say that being is or that being is being is to take the happening of being for some*thing* that is, and hence for its effect.

The thought that being as being shows itself while withdrawing in *what* is being is the thought that the meaning of being as being shows itself while being lost in the excess of the meaning of being as what is being, as thinghood, as objectivity, as "reification" [*Verdinglichung*], a term by Lukćás, used twice by Heidegger in *Being and Time*.[46] This thought grows dramatically in the 1930s and 40s, the years of the *Contributions to Philosophy*, the intense readings of Nietzsche and Hölderlin, the *Black Notebooks*, the origin of the work of art, and many other writings. But this thought grows dramatically not only because of these readings but above all because of Heidegger's readings of his time, a time of the ecstasy of excess and of the excess of ecstasy,[47] totalitarianism, Hitlerism and Nazism, total mobilization and "unconditional anthropomorphy,"[48] and the war of total destruction and extermination, when even the end of being could be conceived of in the thoughts of the "eschatology of being," as Heidegger formulates.[49] That being is being, that the truth of being appears while withdrawing in the total or planetary mobilization of all beings and of the all of beings in universal reification, this is understood by Heidegger as the risk and the danger of being no longer being.[50] This also means, however, the possibility of a turn in being or a change of being, another being of being itself, that could open the possibility of another civilization, of another site of the moment of being, of another beginning of the beginning itself. For Heidegger, being is being in an aletheiological way, appearing as being while withdrawing in the excess of beings, of thinghood, of machination when metaphysics becomes universal and planetary. It could be formulated thus: being is being while withdrawing as being. This indicates, according to Heidegger, a possible change of being, the possibility of another beginning of the beginning—a turn in being. This is the turning core of Heidegger's thought of the event (*Ereignis*). The event corresponds to the turning movement of being, to its "turning paths,"[51] that appears in its own withdrawal, that is, in its *Enteignis*, for it is in-appropriable by human will.

In the dark years of the 1930s and 40s, Heidegger becomes obsessed with reading the end of the world: the end of the philosophical civilization, the end of the first beginning, which does not cease to end. He reads the end of the world as the excess of ontologization and reification, indeed as the excess of the philosophical excess of taking being for *what* is being, taking the event for a being and for an effect, an excess that reaches its utmost form in modern machination, total mobilization, quantification, planning, calculation, will to power, all of which are for him the excess of the philosophical project of universalization that became universal itself. Heidegger can indeed be considered the first philosopher of globalization even if he prefers the vocabulary of the planetary.[52] In the ecstasy of these excesses, in this end that does not cease to end, all forms of existence, of thought, and of saying are being left behind and no other form of existence, of thinking, and of saying has yet appeared. The times are the times of the end of what he calls "the first beginning," the end of philosophical civilization, times of no longer being able to exist, to think, and to say in and through inherited forms and concepts, despite not yet having any other form or concepts to exist and to think. The times are times of both the destruction of all forms and the absence of other forms, times where the risk that even being ends, this risk emerges as much as the possibility of a change in being, of a turn in being and of being, hence the possibility of another beginning. They are times of transition, times of abeyance. *Being is being while withdrawing as being*: Heidegger grasps this while withdrawing of being as time-space. In section 239 of the *Contributions to Philosophy*, a very obscure attempt to describe this "while withdrawing of being" can be read. Here a description of the "oscillation of beyng itself," of beyng with a "y," is formulated precisely where this oscillation between the end of all forms and the possibility of another form of form itself appears to oscillate.[53] Heidegger also describes the "while-withdrawing" as the swing and pulsating beating of nearness and remoteness, emptiness and bestowal, verve and hesitation, and insists that a thinking of this while-withdrawing, of the oscillation of being itself, of this transitional between or suspension must be itself inceptual, insofar as it must think *with the without*—that is, it must bear the absence—of former forms of thoughts; thus, there is still no other form of thought or of existence. The while-withdrawing is described as an abyssal ground, as "temporal-spatial emptiness, an originary yawning open in hesitant self-withholding."[54] Indeed, Heidegger presents an apocalyptic description of the "while-withdrawing of being," of being as transition, as interplay and between-ness, a description from the perspective of the oscillation between the possibility of the end of being and of the change of being inside being, or "the other beginning" of this very beginning.

From a Thought of the Ecstasy of Being to a Listening to the Whiling of Being

After the war, a change of tone can be identified in Heidegger's work. But this is not all—a change in his philosophical narrative can also be observed insofar as the figure of the "other beginning" becomes silent. It would be too quick a reading to say that it disappears; we can read in later texts about "another destiny,"[55] and of course the emphasis on the question of the event and its retraction and dis- or ex-appropriation, *Ereignis* and *Enteignis*. It is also difficult to deny, however, that the thought of the turn is also turned. Carefully reading the lecture *Die Kehre*, "The Turn," which Heidegger held 1949 in Bremen in the cycle of lectures entitled *Einblick in das was ist* (*Insight into that which is*),[56] it seems the turn of the turn is a turn of attention, a turn of attention to the *while* as such. After the war, the increased uses of verb forms that are more gerundive become quite visible. Heidegger turns the focus to the while and meanwhile, *während*, to the abiding and the whiling, *Verweilen* and *Weilen*, using old words such as *Wahrnis*[57] (guardianship), in which the German word for truth, *Wahrheit*, subtly resounds within the word for while, *während*. He stresses the verbal meaning of the word *Wesen*, which as a verb could be rendered in English as going-on.[58] Turning attention to the while rather than to the withdrawing, Heidegger turns his attention to the presence of the present and envisages what was earlier for him an obstacle for attuning thinking life with being—namely, presence, *Anwesenheit*. The step back, (*der Schritt zurück*) assumed by Heidegger as the proper path of (his) thought describes a "leap into the event of presence," to quote a very adequate formulation by Françoise Dastur.[59]

Listening carefully to the difference in tone in Heidegger's work after the war and following the way he puts increasing accent in the question of "presencing" and of the "whiling" and "abiding" of being, it can be said that a certain self-critique or rather a change in his own conception of ecstatic temporality takes place.[60] Indeed, it is possible to claim that Heidegger comes even closer to what I am calling "gerundive time." The critique he made in the *Contributions* about what he had written in section 70 of *Being and Time* is a clear expression of his turning attention to the *whiling of being*.[61] It can even be argued that the focus and discussions about "time-space" are not an attempt to pay more attention to the issue of spatiality but to the whiling of being, from which a spacing and a timing can take place and take its time.

These thoughts, sketched out during the 1930s and 40s become clearest in the lecture called *Time and Being*, in which Heidegger exposes step by step his further thinking, rather than rethinking, of the turn of being in being. In 1949, Heidegger answers the question "What is to be done?,"[62] the question that framed modernity in terms of revolution, with another question—namely, "what must we think?"—some years later rephrased as "what is called thinking?"[63] He claims we must think the turn at stake in the withdrawal of being in the beings, accomplished in planetary manner in the epoch when and where being appears as *Ge-stell*, as "positionality" or "enframing."

Here some clarity about *Ge-stell*, the positionality of being, which is defined as the essence of technology, is needed. This positionality, *Ge-stell*, is not to be understood merely in the sense that the reified meaning of being is posited everywhere—this might be true but it is nonetheless not the most crucial. The most crucial aspect is that this positionality is itself a turning movement, in which everything leaves behind all ontological positions, being so to speak deontologizied for the sake of becoming reontologizied whenever, wherever, for any purpose whatsoever. This is what Heidegger aimed to show in his notion of *Bestand*, presented in the lecture "The Question concerning Technology" and currently translated as dispositive or resource.[64] We must think, Heidegger insists, about the "turning danger" [*kehrige Gefahr*],[65] about how the forgetfulness of being is turning into guardianship [*Wahrnis*] of the essence of being, that is, of the presencing of being. He even proposes to think Hölderlin's lines "But where danger is, also grows the saving power" [*wo aber Gefahr ist, wächst das Rettende auch*], even more intensely, into the extreme, [*in das Äußerte*].[66] The withdrawal [*Enteignis*] of being is thought as a prelude to the event [*Ereignis*] of being. It is a thought that acknowledges that the very thought of overcoming is to be overcome. Thus, as Heidegger affirms in this lecture: "The only purpose of this lecture was to expose how being is as the event" [*Allein die einzige Absicht dieses Vortrages geht darin, das Sein selbst als das Ereignis*].[67]

In order to think this turn, it is necessary to follow how the turn in being is performed. Such following is only possible if thinking and writing also perform a turn, indeed, if thinking and writing become performative. This performativity is a strong mark of the late Heidegger. At the beginning of *Time and Being*, Heidegger renders explicit this demand when saying that "the point is not to listen to a series of propositions, but rather to follow the movement of showing [*den Gang des Zeigens*]."[68] At stake is the movement of showing inherent to being itself that Heidegger proposes to reproduce somehow in the movement of thinking this showing through a performative thinking-writing.

The movement of showing how the withdrawal is a prelude of the event, how the lack of being is a prelude of another "constellation" or relation between man and being,[69] demands a turn in listening from which a turn in thinking becomes possible. It demands the capacity to listen to how, in the same, another is *already* showing and saying itself. Here a listening to the "same-other" [*Selbander*] to the "phenomenon" of being showing itself in its is-being is decisive, what Heidegger in short fragments written in the 70s, before his death, called *tautóphasis* and even *phenomenóphasis*.[70] In question is the possibility of listening and seeing how being *is* being and becoming attentive to how the "is," the "*ist*" *blitz*, that is, casts light. Indeed, in the recent published self-critical remarks on his own thoughts and writings, we can find Heidegger coining a new verb, "*zu isten*," that could be rendered in English with "to is," in his attempts to rethink the whole project of *Being and Time*.[71]

"The *Ist blitz*, the is casts light"—this phrase is said and repeated in the lecture, a good example of the performative writing and thinking that Heidegger develops more and more in his later texts. What this performance aims to perform is the listening to the is-being-while-being, to lightning-while-lightning, indeed to the whiling of being. It is a question of how to be attuned to the beating oscillation and stretching of the whiling, as if the meanwhile of being were a string that in some points resonates other-than-same in the same. This performed but not described, listening to another sound in the same—is a way of attuning the reading with the while, with the need to "hold on" [*aushalten*] in the presencing that is indeed nothing which one can hold onto. This performative listening-insight into that which is—into the is-being of being—is described by Heidegger as a restrained and sober thinking attitude, that lets presencing presence.

In *Time and Being*, Heidegger thinks the relation between being and time on the basis of a turn: being and time turns into time and being. The turning in the title indicates the question of the turning danger, of the turning paths of the event. He exposes what he has been doing more explicitly after the war—thinking anew—which now means above all—listening anew—the first of all philosophical sentences, the one inherited from the poem of Parmenides: "the same is to think and to be" [*to gar auto noein esti te kai einai*]. Heidegger listens "the same is—(this is) to think and to be." Listening to another rhythm of this inaugural sentence of Greek philosophy, he insists on the difference between *eon*, the present participle of the verb to be used by Parmenides, and *einai*, the infinitive to be which consolidates in philosophical tradition.[72] He pays further attention to the neuter "*to*," "it," as the pronoun of is-being and listens to another meaning

in the neuter, which has historically been the basis of the substan-
tialization of being: he hears the rhythm of the is-being. In doing
so, he addresses implicitly the question of why this historical mode
of thought called philosophy emerged in the Greek language and to
which philologists tend to agree that the possibility of building neuter
substantive forms of verbs characteristic of the Greek language (and
later of the German) enabled philosophy as ontology.

Heidegger hears in being, *Seyn*, the present participle, *seiend*.
He hears in the neuter other nuances. Indeed, already in the period
of *Being and Time*, there is a thought of the neuter of being and of
Dasein that already transforms the mere neutralizing function of the
neuter that, in philosophical language, has formed the grammar of
the substantive meaning of being upon which the empty, abstract,
and logic universality of being is built. In a course held in 1928 titled
Metaphysical Foundations of Logic,[73] Heidegger, somewhat in relation
to Leibniz, revisits some fundamental concepts of *Being and Time* not
only for the sake of meeting certain criticisms, but above all to clarify
some basic notions. Discussing the transcendence of *Dasein* and what
he called "the problem of *Being and Time*," he insists on the "neutrality" of *Dasein*. *Dasein* is said here to be neutral in various senses.
For the first, being neutral *Dasein* emphasizes its nonanthropological,
the philosophical demand to assume *Dasein* as as-structure of radical
nonindifference, insofar as *da-sein*, there-being, is the way of being
which is always concerned with its own being. *Dasein* is neutral because
it is nothing but relationality. *Dasein* is furthermore neutral for not
being gendered; *Dasein* is neither man nor woman; *da-sein* is indeed a
neither-nor structure. As neither-nor, *dasein* is not thereby indifferent
to someone or to each one but "the originary positivity and power
of being."[74] *Dasein*'s neutrality is, moreover, described as the power of
the origin, [*die Mächtigkeit* des *Ursprungs*], and not as the nothingness
of an abstraction.[75] Heidegger also affirms that

> Neutral Dasein is never what exists; Dasein exists in each
> case only in its factical concretion. But neutral Dasein is
> indeed the primal source of intrinsic possibility that springs
> up in every existence and makes it intrinsically possible.
> The analysis always speaks only in Dasein about the Das-
> ein of those existing, but it does not speak to the Dasein
> [being-there] of those who exist; this would be nonsense,
> since one can only speak to those that are existing.[76]

This passage renders clear how close Heidegger already is in *Being and Time* to grasping the gerundive mode of *Dasein* insofar as he assumes clearly a kind of "existential difference" between "existing" and "existence," a difference that does not coincide but corroborates the thesis of the ontological difference between being and beings. In this much-discussed paragraph on the neutrality of *Dasein*, which owes its notoriety to Derrida's critical discussions on the lack of a thematization of sexual difference in Heidegger and the political philosophical implications of this lack, Heidegger also develops the thesis that the neutrality of *Dasein* is to be understood as "the metaphysical isolation of the human being,"[77] as its metaphysical solitude which in turn explains how *da-sein* "harbors the intrinsic possibility for being factically dispersed into bodiliness and thus into sexuality."[78] *Dasein's* neutrality is for Heidegger a way of clarifying *Dasein's* "multiplicity," "strewnness," and "dissemination," as what belongs intrinsically to *dasein*. A "transcendental dissemination" thus belongs to the metaphysical essence of neutral *Dasein*, according to Heidegger. In order to understand his need to develop this description of dasein's neutrality as the basis for grasping its "transcendental dissemination," we shall keep in mind that in these discussions, Heidegger aims to grasp *dasein* from the viewpoint of the fact *that* dasein is existing. In these earlier descriptions of *Dasein's* neutrality, Heidegger shows a nonneutral meaning of the neutrality of *da-sein*. Its main signification is the one of a neither-nor that constitutes the throwness of *da-sein* in the is-being in-being. Though it doesn't seem to have been totally clear to Heidegger himself, the neutrality of *dasein* is discussed in terms of "existing" rather than of "existence."

The attention to the proper temporal mode of exist*ing* as what cannot be grasped by thinking the temporality of *Dasein* as ecstatic becomes however more consistent and pregnant in the later texts and as already observed especially in *Time and Being*. There, Heidegger further deepens the sense of the neutral when "translating" the "is," in the expression "*it is*," *Es ist*, into "it gives," *Es gibt*. Because in the same "is," it casts light on another meaning and relation—it must be said that being gives itself. In the gift of being, what is given is neither a meaning nor something; what is given is the is-being of being. After discussing, in a rhythmic play of sentences, the current understandings of the relation between being and time and of time and being, the aporia from which these current understandings emerge, and further the several epochal changes of the meaning of being, Heidegger, in a very performative textual gesture, says that "while, we were just now thinking about being, we found: what is peculiar to Being, that to which being belongs it remains retained, shows itself in the It gives [es gibt] and its giving and as sending."[79] Listening again and again to "being," "while we thought [. . .] in this way" (in a metaphysical way), the is-being of being gives itself as presencing. In the whiling of being, being gives itself as whiling, as presencing. Heidegger differentiates the presencing—das Anwesende—from the present, Gegenwart; it differentiates the while—während—from which presencing can be thought, from duration. He discusses Aristotle's mistake of seizing the present based on the now and not on presencing. Indeed, the whole lecture can be read as the refutation of a passage of Aristotle from the book on Memory and Recollection that Heidegger does not quote but that reads: "[M]emory cannot remember the present while it is present" [tó de paròn hote párestin]."[80] One could argue that Heidegger is trying to think precisely the contrary, namely, presencing while presencing, on he on. The whiling of being, the presencing, is furthermore differentiated from simultaneity, from zugleich. The closest Heidegger comes to the gerundive meaning of being as presencing is in his concept of Nahheit or nearing nearness, approximating approximation.[81] Further, we should not forget how often in Time and Being Heidegger uses the formula Es gilt, currently translated as "it counts" but which more literally says "it is valid." In the "Es," the neuter of is-being, that is neither being nor not-being, there is a question of accounting, worthiness, value, and dignity, that remains unthought.

Time and Being and other texts from this late period allow us to mark a turn of attention in Heidegger's thinking of the withdrawal into the "while-ness" [während] of being, which demands a thought of time differentiated not only from temporal representation of duration but also from the understanding of time as ecstasy and excess. As stated before, Heidegger uses more and more present participles, such as "während," "weilend," and even "seiend."[82] In the lecture Was ist das die Philosophie? (What Is That Philosophy?), he formulates what until then seemed to have been impossible for him, namely, that Sein ist das Seiende,[83] being is the being, which could also be said with being is being. It seems that he suddenly heard something different in the same word seiende, against which he had been struggling all his life. He listens to its present participle, to its continuous form, to, let us say, its gerundivity. He listened to what is so glaringly obvious in English and what renders impossible the translation into English of the ontological difference between *Sein* and *Seiende*, being and being insofar as it shows what was so close to Heidegger that he could not see it, namely, that *sein ist seiend, that being is being*. At stake in this neuter "*Es*" is the worthiness of the event of being, a worthiness that cannot be grasped by the idea of ecstasy and excess, but demands a thought of the gerundive mode of whiling and abiding.

§⁂

The "approaching reading" of Heidegger's thoughts on ecstatic temporality proposed in this chapter was made in the search for traces of a thought of gerundive time present—even if only nascently—in his philosophical path. In *Being and Time*, Heidegger shakes the foundations of sedimented meanings of existence, proposing an abyssal ontological difference between an infinitive, verbal meaning and a substantive meaning of existence, between "to exist" precisely as a to-exist and "to exist" as existent. The distinction between human existence and existent things, which counts today in a growing body of scholarship as Heidegger's anthropomorphism and anthropocentrism, was read here instead as a strategy to clarify human existence as the comprehension of being an infinitive verb since it is constitutively, that is, in itself, out and beyond itself, in this sense "exilic." This infinitive and indefinite mode of being, which Heidegger describes mostly as "ec-static" and as "exposure," can be easily recognized in the experience of exile. What Heidegger seems to become more and more aware of, after the war, is that rather than the experience of outsideness, there is a sense of nearness, of presencing, that grows from the insight into the whiling and abiding of being, and which thoughts on coming to be and overcoming are not really capable of grasping. Heidegger begins then more and more to pay attention to the gesture of thinking through writing, which is a form of bringing thinking to language and listening to the language of thinking while thinking. Because the scope of the present book is to unfold the *experience* of gerundive time, of being *while* being, of existing *while* existing, insofar as this gerundive experience of time corresponds most intimately and intensively to the experience of time from within exilic existence, the focus has been to follow how this so-called gerundivity appears or rather shimmers in Heidegger's thoughts. Even if the word "exile" is silent, what speaks through these thoughts is an experience of time that, although belonging to everyone, thus it is the very experience of existing, becomes acute in exile, and indeed so acute that one can hardly think it in words.

Chapter 3

Time Absent/Time Present

(In Discussion with Blanchot)

L'écrit ça arrive comme le vent, c'est nu, c'est de l'encre, c'est l'écrit, et ça passe comme rien d'autre ne passe dans la vie, rien de plus, sauf elle, la vie.

—Marguerite Duras

The Flight of Philosophy into Literature

To overcome the need to overcome—this motif pervades Heidegger's thought on serenity, *Gelassenheit*, and the splendor of the simple. Heidegger formulates it in terms of overcoming metaphysics. Metaphysics in its literal and current sense means beyond (*meta*) physics (*ta fusika*), that is, beyond the visible, the sensible, the said, the inherited structure of meanings. In other words, the term "metaphysics" names a search beyond the given, a search undertaken for the sake of grasping and conceiving the ground, reason, or ultimate truth and meaning of all that is, what indeed in the metaphysical vocabulary is to be understood as the beingness of all beings. When he speaks of the need to overcome metaphysics, what is to be overcome is also a need, namely, the need for this search for grounds and reasons. This search is for him not due to the distinction between being and nonbeing, but rather to the oblivion of the distinction between being and beings, which results in the forgetfulness of being itself. This search for grounds and reasons has a name: philosophy. To overcome metaphysics is, in Heidegger, entirely connected to his views on the need to overcome philosophy, the need to overcome the need for a search of the beyond. Therefore, the movement of overcoming over-coming can be understood as the search for a beyond of the beyond because metaphysics—beyond physics or beyond nature—is the spirit, and spirit is language, thought; indeed, what determines the human, to overcome metaphysics means to overcome the human. More specifically, overcoming overcoming denotes the need to take a step beyond a certain meaning of the human in order to better understand what it is that provokes the very need to step beyond. To overcome metaphysics would then mean to step beyond every stepping-beyond. The "issueless," an expression that comes from Beckett, names well the impossibility to exit such thinking.[1] It is the "issueless" of repetition and tautology. Heidegger turns the question of overcoming, *Überwindung*, into a question of *Verwindung*, in order to denote another movement in thought. *Verwindung* is a difficult word to translate. It has been translated in various ways into English, most recently as "conversion" or converting,[2] not in a religious sense, but in the sense of remaining and enduring the turning point while turning, that is, enduring or holding on (in suffering, for instance). *Verwindung* implies a claim to remain where one is, in the is-being, in the no-way-out of being itself. The question that we are pursuing persists in these thoughts: How to think the is-being? How to say being while it is being? How to think and say it, if to think and to say is to step beyond it? What remains to be thought and said, then, is not the site from which or to which the "stepping beyond" of thought and language steps, but rather to think and say the stepping beyond in its own movement. This is a crucial challenge in any attempt to conceive of and grasp the experience of time in exile. Thus, from within exile time does not really pass; time remains in the meanwhile, moving without move-ment the whole past and the whole future. In exile, time emerges as the tension of between, in which a sense of presencing presents itself in its outmost unbearable weight. In the movement inherent to the temporality of the meanwhile, the beyond that belongs to every stepping does not step beyond, remaining out of itself in itself. What remains is therefore to remain within this beyond-ness, or better said, within this outsideness from which no one can escape.

Considering the long path of Heidegger's thought, one could claim he goes from the formulation of ecstatic and enrapturing time as "being in itself outside of itself," to a formulation of time-space as "being inside the outsidedness of being," in his late work—a slight but significant alteration, one that brings us closer to a thought of gerundive time. This later thought is the core of his concept of intimacy, *Innigkeit*, and insistency, *Inständigkeit*, in being. Heidegger's thought on the intimacy and insistency of being must be dissociated from any thought of interiority. Meant here is intense nearness to being, understood experientially as a verb, which in Heidegger's late vocabulary is said with "presencing" (*Anwesende*) and hence as what has no interior; thus, as verb, being is nothing but being. To think and say the intimacy and insistency in being is rather an attempt to think and say being while being. Such an affirmation, however, only makes sense if being is not grasped as something outside thinking and saying. In question is how to think and say the thinking while thinking, to say the saying while saying—and as such to experience being while being. But how, if to think the thinking and to say the saying steps beyond thinking and saying, and turns them into figures of thought and words of language? The question of being qua *presencing* is the question of a demand or a call not only to think the thinking and say the saying, but above all a demand to think and say the withdrawal of thinking in thoughts and the withdrawal of saying in the said. But how? Here thinking and saying encounter their limit. Heidegger's late thoughts are thoughts at the limit of philosophy. At this limit, philosophy touches literature, though not, as one might expect, solely poetry.

In 1952 Heidegger held the famous series of lectures entitled *What Is Called Thinking?* (*Was heißt denken?*),[3] which he read aloud on the radio. In these lectures, we find a passage in which Heidegger praises Socrates as "the purest thinker of the West."[4] The purity of Socrates, for Heidegger, lies in the fact that Socrates was able to place and maintain himself in the nearness of the appeal of that, which in withdrawing, draws us to thinking. For Heidegger, this is the reason why Socrates did not write anything. Heidegger, the graphomaniac, praises Socrates for being a-graphic, for not writing, because, not writing, Socrates could remain closer to what withdrawing draws one to think: that is, he could thus remain closer to being. Heidegger makes a surprising claim in this passage: he considers that to write is to flee from the draft of this withdrawal and that, after Socrates, all thinkers "with all their greatness have to be such fugitives."[5] Philosophers as fugitives! Philosophers as those who seem to be capable to hold outside of the whirlwind when fixing the movement of thinking in written thoughts and words. And not only that: he also claims that "thinking has entered into literature, and literature has decided the fate of Western science, which, by the way of the doctrine of Middle Ages, became the *scientia* of modern times."[6] This passage can be brought into an interesting relation to Husserl's discussions about the origin of geometry, not to mention Derrida's comments thereon and his thoughts on arche-writing.[7] Instead of following Derrida's reading of this passage as an indication of Heidegger's understanding of being as archi-writing or as the trace of traces, I would like to stress the indication Heidegger gives here about the relation between philosophy and literature based on the claim that philosophy is a flight from being, which finds a refuge in literature.

Philosophy as flight from the blinding light of the given is an old trope in Western tradition. However, the Heideggerian figure of the fugitive philosopher is quite different; in fleeing from the withdrawing of being, the fugitive philosopher flees into literature. In this passage Heidegger understands literature as the realm of the *written* word. The written word presents a refuge; the refuge of a settled and firmed word as the asylum of an inscribed word, which is the refuge from flight. The written word and its world of fixity enable a certain mode of knowledge, a certain epistemic frame that has settled a whole culture and civilization under the sign of science. Here we could read between the lines the claim that thinking cannot think when written, that is, that literature, in the meaning of not-poetry, does not think, an implicit claim that recalls a whole tradition of refuting literature as lie and fiction. In these lectures, however, on *What Is Called Thinking?* Heidegger also presents a thought on the handicraft of thinking, which proposes an opposite figure; the figure of the thinking hand. But what is the thinking hand if not the writing hand? Maybe we could read between the lines of these lectures about a "writing difference" between the written word and the writing hand, between the *written and the writing*. Indeed, writing remained a tacit but nonetheless highly present question in Heidegger's thought.

The thought of philosophy as a flight from the draft of the withdrawal of being can be further developed in terms of the difficulty for philosophy to think the thinking while thinking, to say the saying while being said. This difficulty is the same as dancing on a tightrope, a figure that inhabits not only Nietzsche's thoughts on thinking but also Paul Klee's motif of the dancer on a tightrope. The difficult thought of this meanwhileness is what I am calling a thought of gerundive time. What is in question is how to think and say the stepping beyond *while* stepping and not before or after the stepping, because both cases would be to go beyond the stepping, beyond the movement. How to think, then, thinking while thinking, how to say the saying while being said if the movement of thinking and of the saying withdraws in thoughts and in the said word, as much as being withdraws in the beings. This is perhaps why the said tends to be sad, and philosophical thought can hardly be dissociated from melancholy.

A question must be raised here: Is the experience of giving itself in its own withdrawal the only possible formulation to the unapproachable gerundive time, the time of is-being, the while of whiling of being, which perhaps is nothing but "approaching," is that the only way of seizing the movement of meanwhileness in thoughtful words? Is the gift of being a gift of appearing while disappearing, of showing in withdrawing? Is Heidegger's concept of truth, the aletheiology of truth, still the truth of being when being is grasped while being? In order to respond to these questions, one needs to throw oneself even deeper into the thought and the language of the *withdrawal*—to throw oneself into the vertigo of the withdrawing while withdrawing, trying to follow this thought to its limit, until the point at which it may break down. In doing so, it may become clear that gerundive time in exile is neither ecstatic nor withdrawing, but rather "approaching."

The Literature of the Step [Not] Beyond

It is from the exigency of these questions that a reading of Maurice Blanchot finds its place. As well known, Blanchot's work had an enormous influence on French thought and literature, and it is striking to consider how his thought could be of such significance for such different thinkers as Foucault and Derrida, Malraux and Lacoue-Labarthe;

and we should not forget the deep friendship that Blanchot shared with both Bataille and Levinas. Many topics raised by Blanchot have inspired all these thinkers, as for instance, the "death of the author," the "writing of disaster" as the disaster of writing, "the thought of the outside," and the "neuter." Broadly speaking, there is a *between* philosophy and literature where Blanchot dwelled, and this has been the major force of attraction to his work.[8] Between philosophy and literature, there is *writing*, and that is why one should refer to his work as a work concerned with writing, at most, rather than with literature. Blanchot's literature is more of a "theoretical fiction,"[9] in the sense of a writing that, rather than inventing, *thinks* the writing, and this also marks his novels, if they are still to be called "novels."

Blanchot is perhaps the writer most obsessed with the task of writing down the withdrawal while withdrawing. While withdrawing and disappearing, something unexpected appears. In Blanchot, the experience of withdrawing is deeply connected to the question of exile. In exile, not only the past and the future withdraw and slip away, but also the very experience of being. The experience of exile is apprehended, in very general terms, by Blanchot as the withdrawal of experience, of existence itself, a withdrawal that opens a certain possibility, even if only the possibility of the impossible. Continuously withdrawing, "being" means, in exile, being nonbeing and nonbeing being. This explains his chiastic, "Heraclitan" style, of affirmations immediately negated and negations affirmed, the excessive use of para-doxes, paronomasias and palindromes. Blanchot's "forms according the model "X without X" (to live without living [*vivre sans vivant*]," "to die without death [*mourir sans mort*], death without death, name without name" that Derrida pointed out as the very "stigma" of Blanchot's lan-guage,[10] are expressions of this attempt to write the withdrawal while withdrawing, a writing that aims to erase writing through writing itself (or aims to erase literature in writing). For Blanchot, literature is, to a large extent, writing down this withdrawing which is being itself. It is not Heidegger's sense of withdrawal, which is being withdrawing in the beings, but, and if it can be said that there is a thought of being in Blanchot, being is withdrawing.[11] Withdrawing is a way to say stepping beyond oneself. Being is (the) stepping beyond itself.

As much as Heidegger, albeit very differently—[an abyss binds together and separates philosophy and literature, the German philos-opher and the French writer]—Blanchot is obsessed with the question of stepping beyond the stepping beyond that language and thought themselves are. Thus, thought and language are given as steps beyond the given in a search for the given. This question also concerns the possibility of getting outside the closure of metaphysics, of Western civilizational patterns, a question he shares intensely, despite the divergence of their answers, with Georges Bataille, who also posed it as the search for a way to step beyond capitalism as a totalizing system of reification. There is another aspect of the thought of being as withdrawal that brings together Heidegger and Blanchot, namely, the central figure of a never-ending-end.[12] The two approach it differently, and yet both understand the *while* and meanwhile as an oscillating pendulum *between*, in the no-way-out of an endless end. *Between* is not understood as an interval, but rather as the inexorable immersion in the endless end of a "system" of thought and language, that is of metaphysics, which has become a civilizational model so totalizing that it is impossible either to exist in it or to find a way out from it. According to Blanchot, the question is much more about *existing* inside this tense between than to "face the prospect of being perpetually immobilized, arrested or frozen—stopped dead—in one's track."[13] For Blanchot, on a threshold there is no immobilization, but a tirelessly moving back and forth, a continuous entrance into where one already is, a stepping [not] beyond in this relation. That is why the book *Le Pas au-delà* [*The Step not Beyond*] that is my focus in this chapter could only begin with the phrase "Entrons dans ce rapport" ("Let us enter into this relation"). The book calls "us," the readers, to enter into this relation, the relation we already are and not only are involved with. Heidegger tried to think "between" as a verb, as "*zu zwischen*," if the German idiom would allow such a term to be forged. Blanchot could do it and say it without difficulty, because in French, *entre*—between—is also a conjugation of the verb *entrer*, to enter. "*Entrons*" should indeed be read as, "let us between." Thus, at stake here is neither to begin nor to end, but to between, *entrons*.

In this chapter, I will reflect on the between, *entre* as a verb, proposing a reading that also enters into and stays in the between, a kind of approaching reflection on how, for Blanchot, the temporality of the between, of the meantime, is that of the withdrawing while withdrawing. The temporality of the between is, according to Blanchot, the temporality of the writing itself. The aim of this chapter is to indicate how Blanchot's thoughts on writing *between* philosophy and literature expose a thought of time, which brings him close to what I call gerundive time. This thought of time is a thought of *meantime*, the temporality of the between, the temporality of "this relation." I propose in the following pages to consider how Blanchot's thoughts of the fragment as thoughts of the withdrawing while withdrawing touch the sense of meantime and between. These thoughts are also thoughts of exile and exilic writing; thus, exile is the experience of the withdrawal of being and of the intermittence of time and space. Blanchot's work has been discussed as the work of "fragmentary writing"[14] and as "exilic writing."[15] Even if Blanchot is not an "author of exile," exile becomes an explicit question for him at a certain point—what has been called the "exilic turn" in his work.[16] Differently from Heidegger, who does not speak of exile and focused his thoughts on the distress of up-rootedness, Blanchot recognizes exile not only as human condition or ontological structure of human existence, but also as the fate of persons and peoples, especially of the Jewish people. His thoughts of exile are largely indebted to his relation to Levinas and the latter's philosophy, and they can be read as reflections on the experience of being inside an outsideness without outside. Much attention has been dedicated to Blanchot's thought of the outside—*le dehors*—following the way that exile often is read in terms of "the outside." The question to be addressed here, however, is how to understand the experience of exile from inside, the experience of being "inside" an outsideness without outside. This also entails envisaging exile more in light of temporality than of spatiality. Blanchot's discussions of the fragment, of the fragmentary, and of fragmentation present a clearer explanation of the experience of this being-inside the outsideness, as an experience of a between and of a meantime, which is distinct from any notion of interval, and which exposes a withdrawing that appears as withdrawal while withdrawing. Indeed, in Blanchot's late works, mainly in *The Infinite Conversation*, *The Step not Beyond*, and *The Writing of the Disaster*, it is possible to follow how exile and the fragmentary belong together to such an extent that for Blanchot to exist in exile is to exist as fragment, indeed as fragmentary and as fragmentation. Exilic existence can therefore be described as fragmented existence, and exilic writing should not be dissociated from fragmentary writing.

In the solid studies on Blanchot's "fragmentary writing" by Leslie Hill and on "exilic writing" by Christopher Fynsk, the fragmentary and the exilic are discussed mainly in relation to a future that is somewhat like a promise. "The fragmentary [. . .] is a promise;"[17] "*The Step Not Beyond* in the furtive promise of a form of peace."[18] Bringing together the fragmentary and the exilic, however, while assuming that exilic existence is fragmentary existence, a different reading of the temporality of the fragmentary, which we could call "exilic fragmentation" emerges. In a discussion in *The Infinite Conversation*, of "the fragment word" by René Char, Blanchot connects the fragmentary with exile and "*dépaysement*" (expatriation). Insofar as the fragment is not "the fragmenting of an already existing reality or the moment of a whole still to come,"[19] "being out of one's element does not mean simply a loss of country but also a more authentic manner of residing, a habitless inhabiting; exile is an affirmation of a new relation with the Outside."[20] Relating the fragmentary to exile, Blanchot indicates another sense of existing in the meanwhile that is not a result of the interruption of an already existing reality or of awaiting of a whole to come. Besides of a "new relation with the Outside" and of a "habitless inhabiting," it is also said that the fragmented poem "opens another manner of accomplishment," that it is a "piece of meteor detached from an unknown sky and impossible to connect to anything that can be known."[21] Emphasizing the rigorous distinction between aphorism and fragment, the aphorism as "closed and bounded," as "a form that takes the form of a horizon: its own,"[22] and the fragment as "juxtaposition and interruption," Blanchot conceives of the fragment as "an arrangement at the level of disarray" and as "an immobile becoming."[23] We shall pay special attention to this awareness of the fragment as "immobile becoming," which is here a thought of the meantime and between (as verb, "to between"), which, as indicated, differs from any idea of promise, however "furtive." Indeed, Blanchot quite explicitly says when explaining "fragment speech" in conversation with Nietzsche, "Fragment speech is speech of the between-two,"[24] a between that is "neither the progress of time nor the immobility of a present—a perpetuity that perpetuates nothing, not enduring, not ceasing, the return and the turning aside of an attraction without allure."[25] This connection between the movement proper to the fragment and the neuter indicates a sense of becoming that is completely distinct from any idea of progress, transition, and passage.[26]

Reading in various fragments how Blanchot conceives of the fragment, the fragmentary, and its fragmentation, one encounters this idea of an immobile becoming, "not enduring, not ceasing," a "between-two," which, far from separating or gathering opposites, keeps both in the tension of a neither-nor, neither here nor there, neither before nor after, neither visible nor invisible, neither one nor another, keeps both in a tensioned rope suspended in the air. The figure of the between—in French *entre*—at the core of Blanchot's thinking of the fragment is explicitly related to the thought of the neuter, of the neither-nor. In a passage from *The Step Not Beyond*, he says:

> Enter/between: enter/between/neuter/not being. Play, play without the happiness of the playing, with the residue of a letter that would call the night by the lure of a negative presence. The night radiates the night to the very neuter in which it extinguishes itself.[27]

As previously noted, in French, *entre* is a conjugated form of the verb *entrer*, to enter. Moreover, neuter, *neutre* in French sounds almost like *n'être* meaning nonbeing; it is almost the anagrammatic form of neuter, in French *neutre*, except for one letter, the vowel "u." Through this letter, Blanchot brings in the night, *la nuit*, for him the negative of presence.

In *The Infinite Conversation*, where Blanchot's reflections on the fragment are made very much in connection to the neuter, he also takes recourse to the unusual typographic mark ± ±, not only in order to reproduce the idea of neither affirmation nor negation, neither positive nor negative, but above all to stress the hyphen operating in the neither-nor and as the insight of the neither-nor. It underlines that Blanchot's immobile becoming must be distinguished from overcoming and stepping beyond. Fragment, the fragmentary, and fragmentation—and for Blanchot, exile—shall not be understood as a moment in a whole to come, nor as discontinuity in a continual process, and even less as a stepping-beyond the closure of totality and unity. It means another sense of difference and separation, a difference without contradiction or opposition, a difference of a "relation without relation," an "indestructible disappearance," a withdrawing appearing as withdrawing while withdrawing, indeed, the "disaster" of a stepping beyond that does not step beyond.

We can see this also in the way Blanchot distinguishes his notion of fragment from the romantic poetics and aesthetics of the fragment. At stake here are not opposed concepts that can be compared, separated, or overcome, but rather the following up of the withdrawing of a certain understanding precisely when it becomes a concept, a solid figure. According to Blanchot, the problem with the romantic thought of the fragment, as presented, for instance, by Schlegel, lies in the way it consolidated and thereby became itself a "closed and bounded" figure, like a "porcupine," an aphorism. The romantics made of the fragment an aphorism, a closed sentence bounded to itself. Blanchot turns the fragment back to the movement it inaugurates, back to its own fragmentation. The fragment is neither a totality beside a broken totality—as a soul beside the soul, nor a moment of another whole to come. If we follow here the art historian Jean-Marie Pontévia's "detached thoughts," as he preferred to call his own fragments dedicated to Blanchot, "the fragment" in Blanchot should be read together with the notions of "supplement," "mise an abyme," "heterogeneity," "informal," "parergon," "disintegration," "an-archy," "unfinished," "hybrid" for they are all "resistant rests" vis-à-vis the danger of totalization, totalitarianism, and totality.[28] The fragment, for Blanchot, is a guardian of an absent meaning, not because there once was a meaning present which has now disappeared, or because this meaning, which never was present, is still to appear, but because this meaning is the withdrawing itself. Blanchot speaks of "what remains without remains,"[29] and he calls it "the fragmentary." In this discussion, he introduces what could be called a fragmentary difference between the fragment and the fragmentary. This difference shall not be taken as an opposition, as we are accustomed to understand difference. It aims to indicate the inherent movement of the fragment, the way it fragments, remaining without remains. In certain passages, we can read Blanchot saying: "The fragmentary expresses itself perhaps best in a language which does not acknowledge it. Fragmentary: meaning neither the fragment, as part of a whole, nor the fragmentary in itself." The fragmentary would mean "the infinite continuity of the fragmentary," as Leslie Hill phrased it,[30] a rest that does not leave behind any rest. Through this fragmentary difference, however, the very notion of fragment also changes. The fragment appears as fragmentary "beyond fracturing, or bursting, the patience of pure impatience, the little by little suddenly."[31] The most decisive here is the suddenness in which and through which the fragmentary accomplishes the writing-down of the withdrawal while withdrawing. In this sense, "fragmentary writing" means the withdrawing appearing as withdrawing while withdrawing. It is neither closed nor open but "infinite dispersal,"[32] neither bounded nor unbounded but "connected to the disaster," emerging as nothing but the risk. In *The Writing of the Disaster*, Blanchot defines fragmentary writing not only in sentences about it but in the very way his writing fragments the coherent theory of the fragments, letting the fragmentary break through, thus as much as philosophy and the writing, the fragmentary "is" not but can only become. Fragmentary writing is both "infinite dispersal" and "not even dispersal as system." It is rather "the pulling to pieces (the tearing) of that which never has preexisted (really or ideally) as a whole, nor can it ever be reassembled in any future presence whatever,"[33] the work of the absence of work, and the "energy of disappearing: a repetitive energy, the limit that bears upon limitation."[34] Fragmentary writing is connected to the disaster, being "the writing of the disaster" when disaster is experienced as "stress upon minutiae, sovereignty of the accidental,"[35] as its "imminence," coming upon us precisely when not coming to us.[36] The "immobile becoming" that defines the meantime and the between-two, and that the disaster renders inexorably and overwhelmingly explicit has for Blanchot the temporality of the "little by little suddenly," of the little by little of the imminent. It is not ecstatic as for Heidegger, but imminent, impending. And insofar as the fragmentary lies in the core of the experience of exile, it can be said that in Blanchot's views, exilic existence is imminent and impending existence, indeed existence in the imminent and impending, and in this sense, it is the "infinite continuity of the fragmentary." To the question why exile should be thought in connection to the neuter, that is, to a neither-nor—neither before nor after, neither there nor here, neither the self nor the other—it can be said that from within exilic experience of existing inside an outsideness without outside, what speaks loudly in this experience is the hyphen, the between and meantime of the hyphen that gathers and separates neither-nor. Blanchot brings to light this, let us call it "exilic hyphen," between and meantime, the tension of the imminent; indeed, imminence becoming imminent.

ॐ

How to describe the tension of the imminent? How to describe the temporality of "immobile becoming"? In Blanchot, the tension of the imminent is often treated as "waiting" (*attente*) and "forgetfulness" (*oubli*). It is not ecstatic insofar as it is not assumed as stepping beyond itself in itself. Nor is it time-space in the sense of stepping in the beyond of the outside. In his search for a formulation for the immobile becoming of the meantime and the between-two, we find in Blanchot, a stepping beyond that does not step beyond. It is a stepping not beyond. Blanchot's fragmentary exilic writing can in this sense be called a literature of the step not beyond.

Since Blanchot's first novel, *Thomas l'Obscur* [*Thomas the Obscure*], from 1941[37] and the first collection of his "theoretical fictions" or essays, *Faux Pas* [*False Steps*], from 1943,[38] the motif of "the step," *Le pas*, *the step* into the between, *the step* in between, has been very decisive in his work. His book *Pas au-delà* [*The Step Not Beyond*], from 1973,[39] as much as *L'écriture du désastre* [*The Writing of the Disaster*], from 1980, and the late short piece, *L'Instant de ma mort* [*The Instant of My Death*], first published 1994,[40] all confirm the central motif of the *step*. Step, in French, is *pas*. But *pas* also means "not," used as negation in a very particular form in French, either in *ne . . . pas* or simply *pas*. The title, *The Step Not Beyond* [*Pas au-delà*], sums up in a very concise manner this double meaning of *pas* as the step and the not: therefore, *pas au-delà* is to be read at the same time as "not beyond" and "step beyond." Step beyond not stepping beyond—or not stepping beyond stepping beyond: this is what the title says. Even more, it formulates concisely the whole of Blanchot's work, of Blanchot's writing, as the writing of disaster in the disaster of fragmentary exilic writing. Because, for Blanchot, exile is the experience of stepping not beyond. In writing, the question of stepping beyond the stepping beyond is embodied not in the sense that some thoughts about overcoming and stepping beyond or not are written down, presented or formulated, but in the sense that in writing down "*pas au-delà*," "the step not beyond," this question is presented both in its closure—stepping (not) beyond the stepping beyond—and at the same time in the sense that the closure is itself an opening up of the closure itself. Accordingly, we should put the "not" in brackets [not] and write "step [not] beyond" for the sake of retaining in English its double meaning in the original French.

The duplicity and ambiguity of this "*pas*"—"step" and "negation"—inscribed in the title *Pas au-delà*, *the Step* [*Not*] *Beyond* indicates the Blanchotian way to approach, but also to neglect what I am trying to think in terms of "gerundive time," the temporality of the is-being as time in exile. In Blanchot, one can find a thought that is *almost* touching upon and touched by the whiling of being. This thought is about the experience of a beyond not beyond, as the between of an endless end. It inhabits the between, understood as a no-way-out of the outside, inhabiting the disaster of the no-way-out of the outside, an "immobile becoming," which is the disaster of facing an end that does not end. For Blanchot to dwell, to remain, in French "*demeurer*," to uphold almost as a dying (*mourir*) in this between—means to write, to write thoughts on writing "in the instant of one's death."

Reading attentively the passages in the book *The Step Not Beyond*, where Blanchot names and discusses the expression "the step not beyond," one finds a thought on the temporality of the immobile becoming, of a between-two and meantime as a step [not] beyond. At the beginning of the book, one reads: "time, time: the step [not] beyond . . ."[41] Time thus repeated twice is the step [not] beyond. As a kind of first definition, then, this passage says that the step [not] beyond is time, time, that would lead "outside of time" in time:

> Time, time: the step [not] beyond that is not accomplished in time would lead outside of time. Without this outside being intemporal, but there where time would fall, fragile fall, according to this "outside of time in time" towards which writing would attract us, were we allowed, having disappeared from ourselves, to write within the secret of the ancient fear.[42]

A "time outside of time in time," is for Blanchot the time of a repetition—"time, time"—and repetition, in turn, is what writing repeatedly performs. Later on in the book, Blanchot defines writing as what "repeats itself endlessly." This endless repetition "does not belong to duration," he insists, and it "separates us from any appropriateness as from any relation to an I."

> [S]omething that would be the "step not/beyond that does not belong to duration, that repeats itself endlessly and that separates us (witnesses to what escapes witnessing) from any appropriateness as from any relation to an I, subject of a Law."[43]

As remaining in the between, indeed, in the step [not] beyond—writing is a temporal experience that does not belong to duration and that escapes subjectivity. Some readers of Blanchot would argue that writing means "infinite finitude"[44] and as such "exile from time,"[45] a kind of eternity in time. Writing challenges duration for it enacts a sense of the eternal which is nevertheless very distinct from the eternity of the concept or of the idea. The eternity of repetition performed in writing is rather an arrest, and in the language of Blanchot it could be called an arrest of time. This arrest can be understood as a restless relation to the "meantime, never finished, still lasting, inhuman and monstrous,"[46] recalling some words by Levinas in the essay "Reality and its shadow,"[47] written as a sort of response to Blanchot's text "Literature and the Right to Death."[48] Arrest without duration captures a first temporal feature of the fragmentary and of exilic writing, the one that does [not] step beyond.

Furthermore, writing challenges subjectivity, which is understood as "the consciousness of the self in its identity," to use Foucault's succinct definition.[49] This claim made by Foucault in his reading of Blanchot insists that Blanchot experiences writing as "the breakthrough to a language from which the subject is excluded, the bringing to light of a perhaps irremediable incompatibility between the appearing of language in its being and consciousness of the self in its identity."[50] Considering that modern consciousness of time as duration has been thought on the basis of the relation between time and self-consciousness, from the internal sense of the *cogito me cogitare*, I think that I think, that implicitly accompanies all representations, to recall Kant's interpretation of Descartes, Foucault is quite right to suggest that Blanchot proposes an interruption of this interior connection between time and interiority. Time as repetition and time as noninteriority, these flashes of definition signal the direction into which Blanchot grasps the temporality of the step [not] beyond which is the temporality of writing *between* philosophy and literature as well.

"Immobile becoming," "time, time, time outside of time" in time, time of endless repetition, time that separates "us" from a self and from *the* self, time that does not belong to duration, time without subjectivity, time that arrests time in the meantime and between is above all, according to Blanchot, "repetitive energy" which is the "energy of disappearing," "limit of the infinite mortal" recalling expressions from *The Writing of the Disaster*.[51] The energy of the withdrawal is "repetitive," because it remains, echoing, and it is as repetition and return, these key words of the fragmentary and exilic experience, that a detour through "the stress of the minutiae," through disaster, is eternally taking place. In Blanchot's writing, fragmentary exilic temporality is the temporality of an "eternal detour."

Eternal detour is an expression that Blanchot coins listening to Nietzsche's formulation of the law of the eternal return. Interpreting the eternal retour as eternal detour, Blanchot connects explicitly the experience of "endless repetition" of the end, performed in and as writing to the topic of exile. In Blanchot, there is a thought of time in exile, but this topic is discussed on the basis of how exile and writing are intimately intertwined, and how exilic time is for him the temporality of writing. Writing is therefore the exilic experience *par excellence*, and exile, existence inside the outside, indeed insideness in outsideness, a thought deeply indebted to Levinas's understanding of exile as "the trembling of an inside-out,"[52] is existence continuously stepping beyond not beyond. Indeed, one of the most central passages of the book discussed here, *The Step [Not] Beyond*, a book composed entirely of *passages* [not to forget that "*passage*" in French comes from "*pas*," "step," which allows us to understand here passages, *passages*, as stepping] is the one that relates exile to the return of the detour of exile:

> The "re" of return inscribes like the "ex," opening of every exteriority: as if the return, far from putting an end to it, marked the exile, the beginning in its re-beginning of the exodus. To come again would be to come to ex-center oneself anew, to wander. Only the nomadic affirmation remains.[53]

The return begins anew the exodus; the return is what exposes the proper movement of ex-centration of exile, of a detour. Blanchot considers the "ex" of exile, the opening of every exteriority, the "ex" of excess as a "re," the "re of return," the "re" of repetition that is more of a detour. It is a question why the "ex" of exile should be understood rather as "re," if writing, as he says in the beginning of the book:

> no longer allows you this relation to the being—understood in the first place as tradition, order, certainty, truth, any form of taking root—that you received one day from the past of the world.[54]

The "re" is not to be grasped as the mere coming back to the same as in tradition, order, certainty, law, truth or any form of taking root. It is not the "re" of mere self-referentiality. The meaning of return as a form of "taking root" is the proper dialectic mark of modernity, which explains to a certain extent why modernity understands and names itself with words formed by the prefix "re:" renaissance, reform, representation, reference, revolution, resentment, etc. Thus, the "re" indicates a double movement: on the one hand, it acknowledges modernity as a time of up-rootedness, of loss and lack of roots, a central motif in Heidegger's critical readings of modernity and planetary technology, but precisely therefore also as times mobilized to take root on new grounds and fundaments. And on the other, the "re" unveils that every attempt to [re-]take roots uproots and ungrounds. What is being conceived of in the modern use of the Latin prefix "re" is indeed the closure, self-referential and self-grounding movement of modernity as a constant up-rootedness and taking roots anew, a total mobilizing movement that turns all roots upside-down for reinforcing and grounding anew the need for grounds, roots, and fundaments. This dialectic of up-rootedness and taking roots implies exile and migration, but a certain kind of exile and migration in which the utopic nostalgia and the nostalgic utopia of a return are firmed and affirmed.[55]

Different from this kind of nostalgic and utopian experience of exile and migration, the experience of an exile without the possibility of return, an ex-centric exile, the constitutive return of exilic "detour," is what Blanchot considers to be the core experience of the exodus of the Jewish people. For him, exile names fundamentally the experience of departing and leaving behind, and the central question concerning the Jewish experience, which is the one of expulsion and dissemination, remains in the background. The return and the "re" that inscribes exile as the "opening of an exteriority" is the re of a "re-beginning of the exodus," a return that detours from returning to any form of taking root and hence a detour from return. Thus, what is most decisive in the experience of exile that he has in mind is an ex-centration that cannot be undone, being a return without return. If the restoration of belonging belongs to the desire for return in the exile that marks the Jewish experience, which for Blanchot is the experience that exposes the very exigency of writing, it is the impossibility of a return that becomes destiny. Connecting this thought with his reading of Kafka in *The Space of Literature*, it becomes clear that Blanchot's thoughts

on the exile of writing is deeply indebted to Levinas, from whom he
learned his views on Jewishness and Jewish "nomadic truth," that is,
how Judaism is defined as a teaching of exile and assumes the Jew as
the figure of exile, the figure for "the very possibility of erring, of going
all the way to the end of error, of nearing its limit, of transforming
wayfaring without any goal into the certitude of the goal, without any
way there."[56] To the exilic condition of a return without return, which
is the return of exilic detour, corresponds an exigency of writing. This
exigency is as much the writing of this exigency, performing in writing
the temporality of "immobile becoming," of an eternal meantime and
between, "the incessant intermittence."[57]

The exigency of writing is that of writing "as a question of
writing, question that bears the writing that bears the question."[58] In
the self-enclosing, self-referential, and self-grounding movement of
modernity expressed continuously in its proliferation of "re-" words,
representation and representations attempt to replace being. This is
possible insofar as being is grasped as presence-at-hand for control
and appropriation. The eternal re-tour of the detour of writing, its
endless repetition, offers a detour within the modern enframing frame
of self-referentiality—a detour that renders presence empty. The writing
leads to an "empty space whose void [. . .] in no way prevented the
turns and detour of a very long process."[59] Thus, the certainty of writing
puts all certainty between parentheses. Recurrent figures of writing in
Blanchot appears in this passage: "To write as a question of writing,
question that bears the writing that bears the question," certainty
putting certainty in parentheses, the writing that renders empty all
attempts to represent presence, to say being. However, this can only
be done, by saying the saying and writing the question of writing.

The impression of dealing with empty pages, with the "fair bit
of verbiage" and excess of paradoxical and chiastic formulations in the
abundant work by Blanchot, is not entirely misguiding.[60] The detouring
return, which in Blanchot can be read as a law, the "law of return,"
is largely performative. We find so many repetitive formulas—return
without return, death without death, the glissando of words, paronomasias, attempts, etymological temptations, the contradicting statements: "Wordless in the midst of words,"[61] "to order is not to speak:
nor is to regulate, Language is not an order. Speaking is an attempt
(a temptation) to leave this order, the order of language: even if it is
by enclosing itself in it,"[62] proverbial phrases such as "if to live is to

lose, we understand why it would be almost laughable to lose life,"[63] and uncountable paradoxical formulae such as "the more he encloses himself, the more he says that he belongs to the Outside." All these recurrent formal devices can be read as a literal reading of the law of eternal return, and as a repetition of the same performing of how writing, which is a thing that cannot be undone, is written in the attempt to undo this law of not being able or allowed to undo itself. Thus, "what is left for you to do: to undo yourself in this nothing that you do."[64] It is a form of writing the withdrawing while with-drawing, which introduces an opening in the closure of the tension of a never-ending end. Blanchot is very aware of the sovereignty of the nothing, and if nothing is changed, "the overwhelming overturning of nothing," ("*le bouleversement de rien*"),[65] is, in its repetitive energy, what accomplishes a detour through minutiae. To write as a question of writing instead of a question of what you write about, is pursued here to the point of being besieged in an obsession [in Latin *obsessio* also means besiegement] of a writing that would be capable of effacing itself through the very writing. The return is therefore to be under-stood as re-turning and over-turning, as turning again and again the written, as returning repetition, indeed as endless re-writing. Another quote can be recalled in this context:

> to write is perhaps to not write in rewriting—to efface (in writing over that which is not yet written and that rewriting not only covers over, but restores obliquely in covering it over, in making us think that there was something before, a first version (a detour) or, worse, an original text, engaging us thus in the process of the illusion of infinite deciphering.[66]

A writing that effaces, the "work of the absence of the work," a production that produces nothing," which distinguishes itself by deleting from itself all distinguishing marks,"[67] all these formulas are repeated again and again, so that the writing becomes indeed empty and a-voided precisely through its obsession and excess. Writing as trace and as marks, as a source of history, as metaphor and support for the work of memory, as the civilizational path to bind the pres-ent to the past and to bring the past to a future, is thus incessantly put into question—as if it were possible to write in such a way that the very mark of history would unmark history, to "erase, by traces, all traces."[68] The "disaster" is on the side of oblivion.[69] This endless rewriting, the only one that would be able "to erase, by traces, all traces" is described as "writing according to the fragmentary," "writing at the level of the incessant murmur."[70] This means "to expose oneself to the decision of a lack that marks itself only by a surplus without place, impossible to put in place, to distribute in the space of thoughts, words and books."[71] According to the fragmentary, writing is re-writing, undoing, erasing, effacing. At the level of the incessant murmur, it is returning, repeating and turning back and forth. It is a matter of writing at the limit of writing; thus, "everything is played out in the difference of these repeated terms."[72] The difference that shimmers may be in furtive hesitation in the midst of repetition, just writing "to write at the limit of writing," and this indicates how repetition repeats above all itself, yet "without being able to be spoken in such a way that it repeats itself in the present."[73]

Most central in all those formalisms, formulations and formulae—and Blanchot is to a great extent a formalist and a very literary writer—is indeed the relation between writing and presence, let us say the relation between writing and being, a relation that complicates the struggle between language and being. At stake is how the being of writing, the being of language relates itself to the relation between writing and being, between language and being, between language and presence.

Neither Absence nor Presence— or the Neutral Time of the Between

Blanchot admits of an obscure combat between language and presence, always already won by presence, even if only as the presence of language.[74] Language fights for presence—aims to reach the thing by naming, but in naming it, language loses, kills, sacrifices the thing it names. This is one of the main thoughts already formulated in Blanchot's earlier essay "Literature and the Right to Death,"[75] first published 1948 in *Critique*, where he, in an explicit discussion of Hegel's figure of the negation of negation, lays out his notion of how language presents the absence of presence, of how death speaks in the speaking, of how literary language is inoperative, entirely opposed to production and action, and how the materiality of language renders more dramatic the absence of things and existents, exposing a doubling of the world. Language is the approaching that distances and the only distance able to approach things and their presences. Language absents things, and writing performs the drama of the combat between language and presence, in which there is neither absence nor presence, except the very "there is," the "*il y a*," neither the presence of things nor their absences, but the "there is" of things, opposed to what something is and to the world of predications, definitions, and conceptualizations. The "combat" between language and presence is dramatized in writing, and it is through writing that language, which is struggling for presence, treacherously destroys presence. By rendering absences present and presences absent, writing alienates presence, and makes presence itself alienating, that is, appearing as something other than a presence. However, it is only by remaining in this untiring self-effacement and self-erasing, in the nonway of this outsideness of language that writing, that literary language discovers a "chance" and a "grace." Only before its own death, life can discover its chance and grace in dying.

In these lines of Blanchot's early thoughts that continuously return in his late writings, he affirms that "writing is not accomplished in the present, nor does it present (something), nor it presents itself: still less does it represent, except the play with the repetitive that introduces into a game the temporally ungraspable anteriority of the beginning again in relation to any power to begin."[76] Writing neither presents nor represents anything. This can be read as a current statement about literary writing, about literature fighting for the right of its "nomadic affirmation." Literature, that is, a writing that unveils writing what it writes, does not represent or present things. But how do we understand that writing never writes in present and that writing does not present itself? Is not writing always presenting itself as writing, the revelation of time repeating time, of time arresting time, in the meantime of immobile becoming and incessant intermittence? Is not writing always writing in the present even if defined and understood as "the spectral effect of a movement of repetition and return extending back into the past and forwards into the future"?[77] It is an insistent claim of Blanchot's that writing is never written in the present, that the repetition which writing performs cannot be spoken in the present. According to him, one writes to forget one's own name, to become anonymous, to separate from consciousness and its world of interiority and self-identity; in this sense, writing is never accomplished in the present in which time and interiority become synonymous. Moreover, as Blanchot insists in many passages in his later texts, writing is never in the present for it also erases what was never traced before. But the sentence "writing is never accomplished in the present" can be read in another way, namely, in the sense that writing is never accomplished in the *present tense* because it is only accomplished in the tension of the instant; thus, if I write "I write," or even if I write "I am writing," the writing is always already written and remains as what has been written. This is a mode to understand the central affirmation in *The Step [Not] Beyond* that "only the *nomadic* affirmation *remains*."

It is a commonplace in studies and readings of Blanchot to note his a-void-ance of the presence of the present and of the present of presence. But in which tense does writing write? Is there any presence that can remain in the unsettled, "nomadic affirmation," in the mark that unmarks, in the trace that erases all traces, both the traced and the never traced, that is called writing? Asking this question to himself, Blanchot writes: "But what remains of presence when it has to hold on to itself only this language in which it extinguishes itself, fixes itself? Maybe only this question."[78] What remains seems to be nothing but the back and forth of the question, the chiasmic movement, the immobile becoming of the between-two and the meantime, the "passive," meaning the passing that does not pass of incessant intermittence. What remains in this "eternal indecision" is the exclusion of the "instant of the presence," "*l'instant de la présence*"[79]; thus, in writing, the future is always already past and the past always still to come. Writing is defined as "excluding any present mode from time."[80] How do we understand this exclusion of any present mode? The intensive present of the instant excludes the instant of the presence. What remains—which is writing—seems to be an instant without instant, or, it could be said, the "there is," "*il y a*," that is not, neither something nor nothing, both no longer and not yet. The instant of writing cannot be written, and this entails the aporia that writing is never accomplished in the present tense because writing can only write in the tension of the instant. For Blanchot, the question is thus not so much about the fugacity and transiency of the instant, but about the instant as tensioned instance, as a tensioned instance that gives itself as imminent impendency. At stake is therefore the imminent impendency of the instant and of the intensity of the withdrawal it involves. It is the intensity of the loss involved in the tension of the instant that haunts Blanchot. As imminent and impending, the instant is an uncanny instance, neither present nor absent, neither past nor future, a between-two and a meantime. Such meaning of the instant emerges in the "instant of my death."

The Instant of My Death is the title of a very short text, published 1994. It can be read either as a novel or as an autobiographical text, either as a piece of theoretical fiction or even of fictional theory, but also as neither the one nor the other, as a text without literary genre, a narrative stepping [not] beyond theory. In this short narrative, Blanchot, the writer compromised before the war with French fascism and with anti-Semitism, presents a kind of confession of his guilty nonguilt that may seem even to be more of a fictitious confession and testimony.[81] The narrative is about "a young man—a man still young—prevented from dying by death itself." It is about "a man still young" that was almost shot by the Germans and that almost became a victim of Nazism. In this "almost" lies the difference between testimony and confession, between truth and fiction, between autobiography and authorship. Is this text an unconfessed confession of Blanchot, a testimony of his nontestimony, the attempt to un-write even more than to erase what has been written through the rewriting of what could have been written, but precisely therefore never can be written?

In his readings of this piece, Derrida presents a detailed analysis of the "strange position of the narrating ego in this narrative."[82] He includes this narrative mostly in the genre of the autobiography in which fiction and testimony are no longer clearly discerned, as much as the difference between what was experienced and what was not. In the role of a reader of Blanchot and of his narrative in particular, Derrida puts himself in a similar autobiographical position when bringing the testimony of a letter he received from Blanchot 1997, one year before he wrote this sort of commentary for a lecture. Derrida quotes the following line of the letter: "July 20. Fifty years ago, I knew the happiness of nearly being shot to death."[83] Through the testimony of the letter, Derrida aims to show how Blanchot's gesture of sending him this letter attempts to testify for the witness, thus challenging what Paul Celan considered to be the limit of testimony, namely the fact that "no one testifies for the witness." In doing so, Derrida unveils the possibility that Blanchot wrote a narrative of what could have happened because it should have happened if what happened did not happen—that is, if he had not previously been involved with fascism and not indeed made anti-Semitic statements. Thus, this short narrative allows for the possibility to be read as a narrative of the un-writing and as an attempt to erase what had been written—and therefore could not be erased—through the very writing.

In this piece, several crucial motifs of Blanchot's works are presented and receive a narrative concreteness. The "outside," this central motif of Blanchot, le dehors, receives a concrete tenor. When written as the imperative "get out," cried out by a German, we can behind the French dehors, hear the resonance of the exterminating German heraus, the imperative to exit and face the commando to kill and to die. At the instant in which Blanchot/the narrator was about to be shot, time is suspended; this is the motif of his well-known early novel, Death Sentence, l'Arrêt de Mort, from 1948. It is moreover the motif of both Goya's painting The Third of May 1808 and Manet's painting of the execution of Maxiliam in Mexico.[84] In this frontal experience of the front line that unites and separates life and death, the tensioned imminence of the instant renders impossible the distinction between fiction and testimony, experience and nonexperience. In the eye of the about to happen, of the imminent and impendent, the instance of what will already have taken place, there is no way to draw the line between being and not being, between presence and absence. Death—which is always a sentence—is also always suspended, insofar as it will always come. To confront death is to confront the eternity of noneternity. Facing the instant of "my" death, one is alive and at the same time dead. Facing this instant as the instance of the imminent, the experience is made that there is "no more ecstasy: the feeling that he was only living." At stake is the experience of "only living," being itself a suspension of the ecstasy of life, of history, of existence—in which one is neither alive nor dead, neither one self nor another, neither here nor there. At the instant of one's death, thrown in the instance of the imminent, one is suspended in the neuter, in the "ending and unending agony" of neither-nor.[85]

Because the tensioned instance of the instant is understood on the basis of the experience of existing—which for Blanchot means writing—as being about to die, this narrative can be also read as a piece of autothanatography, as Lacoue-Labarthe has done. He connects it to a long tradition in philosophy and literature from Socrates and Plato to Montaigne, from Rousseau to Mallarmé, who are witnesses of the dying, as the expression for the passing, the stepping beyond the inconsistence of the present.[86] Lacoue-Labarthe insists that Blanchot assumes death as "the categorical imperative of thought, of Literature" and that this short narrative of Blanchot should be read as "a testamentary text,"[87] as the testament of a whole history and civilization, the civilization of the "ending and unending agony" between philosophy and literature.

The suspension implied in this agony, the agony of the neuter— neither alive nor dead—is described in *The Instant of My Death* as the meeting between two deaths: the death outside him and the death that is dying inside him. Here, thoughts on Rilke and the distinction between proper and improper death, between an anonymous and the own death, discussed some years before in *The Space of Literature* from 1955, return with subtle but nonetheless cautious detours.[88] It returns as the meeting or conflict between death and dying, as what could be called a "mortal difference," which Blanchot also describes as "dying: that which does not rely on life; but it is also death that prevents us from dying."[89] Dying, the nonarrival of what comes about, is opposed to death and to the dead. Dying is invisible, death, irreversible. For Blanchot, death is the loss of losing,[90] that is, of dying. At stake in the "instant of my death" is the interplay between or clash of two negativities, the one being death as categorical imperative—of thought and literature—and the other the transcendental dying as a condition of existence itself, as Lacoue-Labarthe pointed out.[91] This clash emerges, moreover, as the intertwining of two times, objective time and fictional time, what is happening and what is about to happen, but in such way that the only happening is "the imminence of what has always already taken place."[92]

"The instant of my death" is the deep experience of the neutrality—neither-nor—of the imminence of the instant. The imminence of the instant can be called time in neutral. Neither something, nor nothing, neither alive, nor dead, neither here, nor there, both a no-longer and a not-yet. Derrida does not let it pass unobserved in his reading how this piece of testimonial fiction, *The Instant of my Death* is related to *The Step [Not] Beyond*, which for him expresses the "logic" of Blanchot's thoughts on the neuter, that is, of the between. He insists that the difficulty in the "logic" of the step [not] beyond is not the philosophical or speculative logic, but the fact of being a logic "beyond all dialectic, of course, but also beyond the negative

grammar that the word neuter, *neuter*, seems to indicate. The neuter
is the experience or passion of a thinking that cannot stop to oppose
the opposition without however overcoming the opposition—neither
this nor that, neither happiness nor unhappiness."[93] Most striking,
however, in the "logic of the step not beyond" is, Derrida continues,
"the event, thus a passion for the experience of what arrives must be
passion, exposure so what one does not see coming and could not
predict, master, calculate or program."[94] As the logic of a passion
for the experience of what arrives, the "logic" of the neuter relates
interrupting the relation with the neuter upon which the logic of
philosophy qua metaphysics is build. Derrida recognizes in Blanchot
a literary thought of the event opposed to the philosophical thought
of being, and the event as "what one does not see coming and could
not predict, master, calculate or program," as imminent impendency.
Reflecting upon the relation without relation between fragmentary
and exilic logic of the neuter and the philosophical neuter, Blanchot
remarks how the Greek pronoun "*to*," meaning both "the" and "it,"
introduced the mark with a sign and "the decision of a new language,
a language later taken over by philosophy at the price of this neuter
that introduces it."[95] Blanchot insists that the neuter that introduces
the language of the neuter in thought withdraws through this introduction, neutralizes itself, and, since the beginning of philosophy,
misses the neuter despite having marked it with a sign. In this long
passage from *The Step [Not] Beyond*, Blanchot takes into account
that at the same time that the neuter has been considered the very
condition of possibility of philosophical language and conceptuality,
understood as the event and passion, the neuter marks a marginal
tradition in Western thought, indeed a thought not only of the margin
but above all *at* the margin, *ad marginem*, which is the thought of
being outside all insistence on the interiority of the essences of beings
and of consciousness and self-consciousness. In its proper marginality,
the neuter is the inclination towards the night,[96] a source for all
forms and possibilities of affirmation and negation.[97] Trying to break
with the logic of affirmation and negation, the *neuter* in Blanchot
takes the instance of the instant rather as an *echo*. A key passage
reads:

> "Which of the two?"—"Neither one nor the other, the
> other, the other, as if the neuter spoke only in an echo,

meanwhile perpetuating the other by the repetition, that
difference, always included in the other, even in the form
of the bad infinite, calls forth endlessly the balancing of a
man's head given over to eternal oscillation."[98]

The return of exilic detour, the back and forth of repetition repeating
itself without being able to speak of it in present is taken as "an echo,
meanwhile perpetuating the other by repetition." The fragmentary and
exilic logic of the neuter proposed by Blanchot is the "logic" of an
echo, more a kind of echo-logy than echolalia, in which affirmation
either of affirmations or of negations makes echo of itself to the point
of dispersion, dispersion going even to the very silence dispersed."[99]
The logic of the neuter proposed by Blanchot differs from the philosophical neuter insofar as it presents the "logic" of dispersion taking
place in the return without return, in the exilic detour of an echo.
Thus, the echo keeps echoing within itself "in order to withdraw itself
from the continuity."[100] Whereas the philosophical logic of the neuter
is the logic of totality and unity—the logic of the "the" and the "it,"
the logic of the "thing itself," of Being as the Being of all beings, the
literary logic of the neuter is the one of dispersion through echoing,
dispersion through "incessant intermittence" and "eternal oscillation."
In the instance of imminence that, according to Blanchot, defines the
instant, dispersion takes place through echoing.

In this chapter, a tentative effort has been made to reflect on
Blanchot's thoughts on exilic temporality, on the temporality of the
experience of time from within the outside without outside that for
him defines exile, as if one were searching for a place with "a finger
on a map," to recall an image by Paul Celan.[101] Instead of insisting
on the figure of promise and waiting, in thoughts of the "to come,"
we have focused on his writing between philosophy and literature,
understood as writing of the agony of neither-nor, of the neuter, a
writing of the between and meantime. In our discussions, a sense of
the instant emerged as imminent impendency, growing less out of what
might happen than through the tireless return of exilic detour, the
incessant intermittence performed in echoing repetition. For Blanchot,
"immobile becoming" defines the writing at the instant of "my" death,
as the writing of disaster, which is a writing searching to write the
movement from thought to language and language to thoughts, the
one ex-centrating in the other, the one exiled in the other.

In Blanchot's writing of thoughts about the temporality of fragmentary and exilic writing, the figures of the between and of meanwhile are understood as "in abeyance," *en instance*, as arrest and suspension of time,[102] eternal oscillation, immobile becoming, incessant intermittence, instance of imminence in which echoing dispersion, dispersion through repetitive echoing takes place. What has always already dispersed, giving itself neither as this nor as that, "also refuses to present itself in simple presence." Because according to Blanchot "simple presence" is and must be refused, the dispersive gift of the fragmentary and exilic let itself be apprehended "only negatively, under the protective veil of the no."[103] If throughout all the pages that can be found in Blanchot's work about the neuter's between and meantime, we read the insistent affirmation that writing neither affirms nor negates, it is because in question it is the interplay or clash of positive and negative, of affirmation and negation, rather than a writing that writes down the coming to writing. Following the thought of gerundive time, which I understand as time *experienced from within* exilic existence, as experience of the meanwhileness from within, what is at stake is not the withdrawing appearing as withdrawing while it withdraws but the naked, simple being while being, not so much the "it is," but what I am trying to express in the odd formulation "is-being." Blanchot's literary thoughts and thoughtful literature, a literature that moves back and forth in repetition of chiastic detours, one misses the "as if for the first time," one misses birth as the experience from within which one is always already as if for the first time. What withdraws this strong writing of and in the withdrawal is gerundive time, the naked is-being, is-existing while existing, the is-writing while writing, the one that "passes like wind," recalling the words by Marguerite Duras, that passes like life itself passes,[104] the immediacy of life leaving dying, in every life and death.

Chapter 4

Time Being

(Reading Gerundive Time with Clarice Lispector)

Where am I going? And the answer is: I'm going.

—Clarice Lispector

Reading Time and the Time of Reading

In the previous chapters, I attempted to show how Heidegger and Blanchot approach the is-being, the gerundive temporality that marks the experience of time in exile. Heidegger does it more explicitly in terms of "presencing" and "whiling," and Blanchot in terms of "effac-ing" and "disappearing." Both come close to what I am calling the is-being, in their discussions about the "neuter" departing from the urgent necessity to leave behind deep-rooted habits of thought and language. In his late work, Heidegger connects the question of whiling and abiding with thoughts on *Gelassenheit*, releasement or serenity,[1] and Blanchot follows the Levinasian path of the thought of passivity. What matters in these central verbs, in the German *lassen* and in the French *passer* (which the word *passif* comes from) is the "*pas*," the step and the not, the step [*not*] beyond. Both Heidegger and Blanchot left behind to us, their readers, the most anguished works of thought and language about the search for leaving and about abandoning the search for a beyond. Both were in search of a step beyond the need for stepping-beyond; both were in search of a thinking and writing

83

of the withdrawing of being itself, the withdrawal of the presence
of the present, assuming, albeit in very different ways and paths of
thoughts and language, that a "not," a "*ne pas,*" belongs originarily
to being itself and to the event, to recall a passage by Heidegger in
the *Contributions to Philosophy* that reads:

To be sure, a "not" does essentially occur in the hesitant with-
holding if grasped more originarily. But that is the primordial "not,"
the one pertaining to being itself and thus to the event.[2]

The experience of leaving, *lassen, quitter, abandoner, passer,* is
a difficult one; as much as the experience of being expulsed and
disseminated, the leaving and the guilt of leaving mark strongly this
difficult experience that is exile. Perhaps, rather than of exile, we
should speak of leaving, or in the language of Osip Mandelstam, we
should speak of "the science of departure."[3] In their attempts, Heidegger and Blanchot remain caught up or entangled within the need
to overcome overcoming, to transcend transcendence, a situation of
anguished absence of exit. They remain captured by the figure of truth
as unconcealment, of appearing in withdrawing, which, it should be
noted, is a narratival figure of the apocalypse. Of course, we should not
forget the differences that distinguish them, but, at the same time, it
is difficult to deny how they share the figure of absence and absenting,
of fading away, and its crepuscular silvery realm of transitivity. Against
the hegemony of presence and its domain of entities and substantial
meanings, Heidegger and Blanchot propose nonetheless a thought of
the "there is," it is, it gives, the "*Es gibt*" in Heidegger and the "*il y a*"
in Blanchot. Both understand the temporality of the neuter as interruption of chronological time, and propose thoughts about time-space,
and about the eternal detour in the return and the eternal return in
detour. Both write and think at the limits of language, of writing, of
thinking, at the limit at which what must and is valid—*es gilt*—to
be said, written, and thought touches what *cannot* be said, written,
or thought. They are both in the space of literature, writing through
thoughts and seeking to transcend the transcendence of writing. They
are both in philosophy, thinking through the written, seeking to transcend the transcendence of thoughts. They are between philosophy
and literature, between theory and literature. They are located in the
suspension of both philosophy and literature: thinking at the "end of
philosophy," writing at the "end of literature" and are very attentive
to the event of thought in language and of the language of thoughts.

Both came close to what I am calling gerundive time; they came to a thought of existing at the edge of existence. In them we certainly find a thinking and a writing about something very close to gerundive time, but still gerundive time remains to be thought and written *while* being thought and written.

ৡ

In order to continue developing this reflection on gerundive time as time in exile, I propose to bring in another voice. This voice writes in the Portuguese of Brazil, a language growing in the estuary of colonization, of different layers of immigrations, of mixtures, of the autochthone becoming heterotochtone and the heterotochtone becoming autochthone. I mean the writing voice of Clarice Lispector, herself a child of exile who arrived in Brazil at the age of two months, before having a language to carry in her arms, to call her own. She considered herself Brazilian, and Portuguese her only language of writing, which is to say, of being. Clarice is one of—if not *the*—greatest Brazilian writer of the twentieth century.[4]

As a way to address the question about the temporal meaning of exile as the experience of gerundive time in the work of Clarice Lispector, I propose an "approaching reading" of some lines of her novel *The Passion according to G. H.*,[5] from 1964, a book written in neuter, as she says herself: "I have no words to express, and speak therefore in neuter."[6] This is a book written from out of a radical leaving-behind of all hope, of abandoning the wish to transcend the is-being. Together with these lines, I will read other lines from *Água Viva* (1973), in which the gerundive writing of Clarice acquires even wider dimensions. How, then, are we to read Clarice Lispector? And how are we to read these lines of her books?

In her seminal readings of Clarice, Hélène Cixous proposes a kind of poetics of "reading according to C. L.," according to Clarice Lispector. Using the paraphrase of Clarice's title, Cixous aims to find in the texts by Clarice the resources to enter into them. On the one hand, she assumes the impossibility of holding Clarice's texts in the hands of theory, insofar as they are as *Água Viva*, which in Portuguese can mean both a spring or a fountain and a jellyfish. In order to read a living water, Clarice's text, a text that is more a jellyfish, one must let the text overwhelm the reader and even develop a certain "capacity of improvisation"[7] through which different modes of reading can be performed, such as singing it, for instance.[8] On the other hand, Cixous describes the act of reading Clarice like being "before the law," as in Kafka's story, in which the fundamental prohibition "You will not go through" repeats relentlessly. The path chosen by Cixous is much more the one of a reading guided by "the problem of the law, the word, writing and the (libidinal) structure of the writer"[9] than the one of improvisation. It is in many aspects a Kafkaesque reading of Clarice. She considers that

> when we read a text, we are either read by the text or we are in the text. Either we tame a text, we ride on it, we roll over it, or we are swallowed up by it, as by a whale. There are thousands of possible relations to a text, and if we are in a nondefensive, nonresisting relationship, we are carried off by the text . . . But then, in order to read, we need to get out of the text . . . At some points we have to disengage ourselves from the text as a living ensemble, in order to study its construction, its techniques, and its texture.[10]

The need to choose beforehand a reading guide, for instance a Kafkian view, to describe the reading as a choice, "either . . . or" ("we are either read by the text or we are in the text"), and to "get out of the text" and to "disengage ourselves from the text," casts the reading of Lispector as an act of deciphering the rules of a construction, and hence of a newly formed form. Cixous's very inspiring readings of Clarice are investigations in the techniques and textures of a living form, attempts to follow her literature as "a trinket of water," another English translation of Água Viva. Cixous assumes that Clarice's text "disobeys all organizing laws, all constructions and that goes very far,"[11] but she nonetheless looks for the order in this disobedience, for the organization of this disorganization. Cixous admits without hesitation that Clarice's literature "is not into a law that represses differences, but into one that formalizes, that gives form"[12] and seeks the laws of this formalization. In her readings, she searches both the internal coherence of Clarice literature and the way she differs from other writers and philosophers. She acknowledges how Clarice's first published novel, *Near to the Wild Heart*, and the late *Água Viva*, for instance, maintain the same Clarice path, although the new form, the Claricean form, seems more accomplished in the late work. Thereby a certain teleological view of literature, the belief that an author "develops" and "enhances" a singular capacity and style is sustained. Comparing *Near to the Wild Heart*, a title taken from James Joyce, with Joyce's *Portrait of the Artist as a Young Man*, or with Blanchot's *The Writing of the Disaster*,[13] bringing all them to the Kafkian subject of the Law under a Derridean light, Cixous sometimes reduces Clarice to a cliché of a woman writer: Clarice is a writer who follows an "organic order" and not a "narrative order"; she is more of a "savage, uncultured"[14] and "naïve" writer than a cultivated one (as Joyce); and she writes from her "unconscious" resources.[15] In those literary judgments, Cixous cultivates the philosophical dualism of body and soul, of nature and culture, of life and form. Thereby, she can hardly recognize, however, that Clarice's writing leaves behind all systems of hope, based on divisions, oppositions and antagonisms, which are systems of forms, indeed the thought of form itself.

There are no guidelines to read Clarice's books except reading her books. And not even another book by her can become the key to the one being read. To the question of how to read Clarice, there is only one possible answer—one must read her books, each one, *as a world that has precisely begun to exist*. To read Clarice, one must experience the exhaustion of any need for interpretative reading, which is continuously requiring to "disengage from the text as a living ensemble, in order to study its construction, its techniques, and its texture."[16] One has to discover that there is no entry to her writing because, enigmatically, one is always already in it. If we insist on comparing Clarice's texts with Kafka's and Blanchot's, then we discover that Kafka and Blanchot present a literature of the prohibition to enter—the law, literature or life—thus there is nothing beyond their doors; in Clarice's literature, however, there is no prohibition to enter but the discovery that there is no entry thus one is already in—literature is necessarily from life toward life. As she remarked once, "[W]hich existence could be previous to its own existence?"[17] That is why the readers have to disarm and dis-form their forms for the sake of discovering that one is already immersed in her writing thus she is being written by life itself. In order to do that, one has to be immersed within the text; that is, one must read it without goals or intentions as one swims in sunny water. The need to develop the capacity of improvisation is decisive here. But even so one has to accept that the meaning of improvisation is also shaken and has to be itself improvised again and again.

How should one read *The Passion according to G. H.*, this book that Clarice herself considered the one that best captured her demands as a writer?[18] Clarice opens the book with a very short note "to possible readers," saying: "This book is like any other book. But I would be happy if it were only read by people whose souls are already formed. Those who know that the approach, of whatever it may be, happens gradually and painstakingly—even passing through the opposite as what is approaches."[19] This is neither a recipe nor a method of reading and interpreting her or the writing; it is not even a request, a claim, or a desire. Like any other book, this book can be read, and read in many ways, but if it would be read by "already formed souls," those who know that the approach of whatever it may be, happens gradually and painstakingly, that would make her—the writer—happy, content. In English, the adverb "already" hides in plain sight the connection to "all read," intimating that one has to be ready for reading, reading in the sense of approaching, "whatever it may be," gradually and painstakingly. How should one read as if one had already read, and how should this "already" be gradual? This and many other enigmatic questions emerge from this reading because approaching "whatever it may be" to read means here to approach without knowing what one is approaching; it means to approach the unknown, to approach an enigma, a mystery. It means to follow "the side of not-knowing," to borrow a formulation by Cixous,[20] but we should add of not-knowing even to not know. And not only that, it means to experience the most difficult demand by Clarice, namely, that "what I am writing to you is not for reading, it is for being."[21] How then does one read not for reading but for being? And how is it that those that have already formed souls would be more desirable as readers when what matters is reading for being and not for reading?

The book is not merely *about* approaching "whatever it may be." The book *is* approaching *the* whatever it may be, and this is what renders it so hard and difficult to read. Thus, what the reading experiences *already* from the start is that what happens in the book happens in the reading, in fact, that what is being written *in* the book is the being read *of* the book. This "whatever it may be" is what both the narrator—meaning the writer and the reader—approaches.

They—both writer and reader—encounter nothing but *it*, the whatever "it" may be. In fact, the story of this book is both the most banal and the most overwhelming. A Brazilian woman, upper-middle class, living at "the top floor of a super structure,"[22] finds herself alone in her apartment. The maid has quit. The Lady enters the maid's room, intending to clean and arrange it, when she is surprised to see how the room has been left clean and arranged, almost empty. She sees her own suitcases near the wardrobe and finds herself somehow depicted in initials "G. H." engraved upon the leather cases. In these initials, "G. H.," she finds her *self* in its nonbeing, finds it no longer a self, because she finds it instead as "what is neither me nor mine." Despite the attempts of some commentators to read the initials "G. H." as an abbreviation for "*gênero humano*," literally human gender, we should be careful not to overinterpret them. They are nothing but marks engraved on leather, as much quotation marks as letters; for the "I," though it is no longer clear whether this refers to the narrator or the narration, is placed in quotation marks: " 'I' always kept a quotation mark to my left and another to my right. Some 'as if it wasn't me' was broader than if it *were*."[23] With the recurrent quotation marks along the text, the written word calls attention to its being written and print character. "I," she, the narrator, G. H. is the writing. It brings the reader to the event of reading, both the reading of the writing while being written and the visual event of noting down this reading writing. In the attention to the being written and read, the focus is no longer on the "I" that writes or on the "eye" that sees. In this attention, the self in its nonbeing is encountered. This encounter happens throughout the whole book. Indeed, the epigraph of the book, written by the art historian Bernard Berenson and quoted in the English original, reads: "A complete life may be one ending in so full identification with the non-self that there is no self to die."

The Lady opens the door of the wardrobe and encounters a cockroach. The cockroach—this is "what it may be," not in the sense that the cockroach decides what it may be, but in the sense that it names precisely the "whatever it may be." The story *is* [the writing of the reading of the writing of] the gradual and painstaking approach of the cockroach—of *the* whatever it may be. Though it is perhaps the only novel after Kafka that has a cockroach as its subject without reproducing anything from Kafka, it is not Clarice's only writing about a Cockroach. In one short story called "The Fifth

Story," she begins by reflecting on the possible titles of a story about encountering a cockroach: "The Statues," "The Murder," or even "How to Kill Cockroaches." She then continues that there could be a fourth narrative inaugurating "a new era at home, it begins as we know: I was complaining about cockroaches"; the fifth story, she then says, could be called "Leibniz and the Transcendence of Love in Polynesia."[24] Whatever title it may receive it is also a variant of the same theme—of the whatever it may be.

The Passion according to G. H surpasses all possible titles and stories about an encounter with a cockroach at the very beginning, opening instead by raising the central question: "But why not let myself be carried away by whatever happens? I would take the holy risk of chance. And I will substitute fate for probability."[25] In Portuguese, the formulations are more complicated. "By whatever happens" translates pelo que for acontecendo, which literally means something like: "by whatever it will be happening." Further on, the Portuguese does not say "I would" but rather "I will have to take the holy risk of chance. And I will substitute fate for probability." The gerund "happening," acontecendo, is conjugated in connection with the future of the subjunctive of the verb "to be" (for). That is why Clarice does not say "I would" but "I will." The future is already happening, and the happening is already continuing and hence in some sense future. This temporal incoherence is part of the profound disorder provoked by the encounter with the cockroach or the whatever it will be happening. The gradual and painstakingly approach of whatever it will be happening approaches the most frightening. What is the most frightening? Being.

> How could I explain that my greatest fear is precisely of: being? And yet there is no other way. How can I explain that my greatest fear is living whatever comes? How to explain that I can't stand seeing, just because life isn't what I thought but something else—as if I know what! Why is seeing such disorganization?[26]

The greatest fear is of to be-being—the cockroach. But there is no other way: the greatest fear is ir vivendo, literally, to will-go living, it is o que for sendo, not only what is being but whatever will-go being.

Beside the insistent and in Portuguese very common combination of the gerund and the future of the subjunctive, in Portuguese, as in Spanish, there are two verbs to say being—ser and estar. Far from opposed, they are used also in intertwined and complex verb forms, as in the expression estar sendo, to be (estar) being (sendo). There are also two forms of the verb "to be" in the continuous form "being," which in English is both a present participle and a gerund. First, there is the substantive ente, formed from the present participle of "to be," deriving from the Latin ens, and equivalent to on in Greek, étant in French, and Seiende in German; then there is the gerundive mode sendo, which is the mode that Clarice uses here and everywhere, and which is very common in Portuguese. The distinction between present participle and gerund is subtle but nonetheless intense. The greatest fear is to live in gerundive time, to go on is-being, so to say. This demands the holy risk of chance, the substitution of fate for probability.

The book is about the encounter with the infinite and dark gaze of the cockroach of the is-being, of the gerund of being. What is scariest is that this gaze, the gaze of the is-being, which she also calls "ancestral life" and the "neutral crafting of life" is looking at her—the writer writing the reading of the writing. The cockroach, the is-being, the ancestral life of life, this prehistorical living resistance to death, looks at "me": is-being sees me. That is the predicament—no longer the familiar predicament of modern consciousness, of "to be or not to be," or of consciousness being able or unable to seize being. How does one approach this gaze of the is-being upon us? One needs to disarm, "de-heroize,"[27] despersonalize, and deshumanize to lose all thought of being, of time, of form, and of humanization, indeed, to abandon all systems of hope. Thus, in order to approach the is-being—the cockroach—ancestral life, life that like a lizard continues, even after being hacked up, to tremble and squirm,[28] one needs to "lean the mouth on living matter." One has to eat the liquid oozing out of the roach of the is-being. One has to eat life and be eaten of life.[29] One has to eat of the forbidden fruit and not be struck down by the orgy of being.[30] Which fruit is the white mass coming out of the roach? "What comes out of the roach is: today, blessed be the fruit of thy womb,"[31] we read. But this, that which comes out of the roach's belly, is not "transcendentable." Indeed, The Passion is the book of wanting the present of the today and not the future.

[I want the present] without dressing it up with a future that redeems it, not even with a hope—until now what hope wanted in me was just to conjure away the present.[32]

[. . .]

But I want much more than that: I want to find the redemption in today, in right now, in the reality that is being, and not in the promise, I want to find joy in this instant—I want the God in whatever comes out of the roach's belly—even if that, in my former human terms, means the worst, and, in human terms, the infernal.[33]

What is being sought here is to "no longer be transcending and remain in the thing itself."[34] And the thing itself is the thing's *is*; it is the neuter of the is-being.

With Clarice, we discover that the thinking of time as either chronology or ecstasy is due to various systems of hope, which, obsessed with dismembering ancestral life, the cockroach, no longer sees how the neutral crafting of life, the roach of the is-being, continues to tremble and squirm, resisting death, despite rational and sentimental strategies to cut it in pieces: "How much I envy you Ulysses [also the name of her dog] because you only remain being."[35] This continuation is also the one that appears when entering "whatever exists between the number one and the number two;"[36] one discovers that "a note exists between two notes of music, between two facts exists a fact, between two grains of sand no matter how close together there exists an interval of space, a sense that exists between senses—in the interstices of primordial matter is the line of mystery."[37] This encounter with the gaze of is-being undermines the privilege of the fear for dying, which is indeed the fear for the future. Hope is the fruit of the fear for the future, of the fear for a future fear. "Hope is the child that hasn't been born yet, a child only promised, and this hurts."[38] *The Passion according to* G. H. does not only rewrite *Being and Time*, but every Book of Hope, even the Bible, rewriting them after having touched the unclean,[39] the cockroach of the is-being, thereby refusing that death is the great unknown. No, death is the only known, it is the future and "is imaginable," she affirms. What remains unimaginable, unknowable is neither death nor dying but only the "right now," [*agora*], the present of the today, the is-being. "But what I'd never experienced was the crash with the moment called 'right now.' Today is demanding me this very day. I had never before known that the time to live also has no word. The time to live, my love, was being so right now that I leaned my mouth on the matter of life."[40]

The right now, the is-being, the instant, the present, this only, is unimaginable and mysteriously unknown. This only, the is-being, with its continuous ancestrality, is the only frightening. "Because a life or a world fully alive has the power of hell."[41] Caught in its gaze, one has no words, except becoming "near to the wild heart"[42] of the is-being; one has to learn to exist (which here means to write) cardio-graphically.

Approaching "the whatever it will be happening" and letting oneself be seen by the gaze, the infinite and dark gaze of a cockroach, one approaches the amorphous or formless is-being. What frightens is precisely the approach to what has no form—the is-being—the on-going being. Throughout the *Passion according to G. H,* and also very much in *Água Viva,*[43] the reading approaches the experience of formlessness, indeed, the experience of the need to leave behind the thought of form. To approach the formless is to exist without hope, and this is what renders the experience of reading this book, a very frightening experience indeed. Because as she writes, "I want disarticulation, only then am I in the world."[44] In this sense, Clarice could enter into conversation with a short fragment by Bataille, published under the title "Formless," in which he affirms, in contrast to "academic men [who can only] be happy if the universe would take shape," that "formless" is not only an adjective having a given meaning, but a term that serves to bring things down in the world."[45] But in Clarice it is even clearer how the internal logic of the thought of form is both revealed and challenged.

In several aspects, it can be argued that philosophy is not only a thought of the being of all beings but also a thought of the form of all forms, of the archi-formal. The philosophical thought of form has a long history, but it was in Kant's critical philosophy that it received a sensuous dimension, through the connection of form with intuition.[46] Kant changed the Aristotelian table of categories in many senses, but a decisive change was to show that space and time are not categories of understanding but *a priori* forms of intuition. Without discussing the connection between form and intuition but rather simply assuming it, Kant opened up a new path in the thought of form and of intuition that was developed both with romanticism and the phenomenological tradition. Phenomenology, particularly the one of Merleau-Ponty, can be considered an attempt to show that space and time are not *a priori* forms of intuitions but rather *a priori* intuitions of form. If, with romanticism and later with phenomenology, a thought of the formless became possible and even necessary to understand the form from the forming process and not the other way around, it is still for the sake of reaching both a form in the forming process and a form of the forming process that the lack of form and its realm of disorder and disorganization received legitimacy. The challenge of Clarice's writing, of the living water of her passion, is that of a writing that leaves behind the search of a form for the forming process. Because for her what matters is the *is-being*, the *is* that the instant is, there is neither form nor forming, but rather formlessness, the "amorphous" matter of life. In writing she tries to "see" what means as much as "to write" in the moment in which she sees, and not to see or write through the memory of having seen in the past instant. Thereby, the fundamental presuppositions for a thought of form as well as for a certain meaning of the forming are shaken. It is the whole meaning of space and time and of being in space and time that is displaced. The gerund of being, the is-being, the will be happening, cannot be grasped as a now-point in a measured succession of before and after on the basis of which different conceptions of time and ideas of forming and process have been defined in the philosophical tradition. Nor can it be apprehended as internal sense as Kant defined time in terms of "motion as an act of the subject (and not as a determination of the object)."[47] The is-being, what "will-go happening," *o que for acontecendo*, the "is," is not an internal subjective sense; in the is-being there is no internal or external, one is already the other, as in a vibrating resonance. A reading of Clarice shows that what resists the thought, or to be more faithful to her vocabulary, what resists the vision of the is-being is not the linear representation of time. A concept of simultaneous time would not resolve the problem. What resists the vision of the is-being is rather the inattention to the draw*ing* of the line, to the drawing of the line while it is being drawn, at the basis of every linear representation. In fact, Kant himself forgets it precisely when calling to attention:

> We cannot cogitate a geometrical line without drawing it
> in thought, nor a circle without describing it, nor represent
> the three dimensions of space without drawing three lines
> from the same point perpendicular to one another. We
> cannot even cogitate time, unless, in drawing a straight line
> (which is to serve as the external figurative representation
> of time), we fix our attention on the act of the synthesis
> of the manifold, whereby we determine successively the
> internal sense, and thus attend also to the succession of
> this determination. Motion as an act of the subject (not
> as a determination of an object).[48]

What resists the vision of is-being is the inattention to the action
and act of drawing, which is forgotten in the attention to what is
being drawn, namely, the line. What blocks the view of the is-being, the drawing of the drawing is paradoxically the very line that is
been drawn. The English word "drawing" is here very illuminating
for its profound ambiguity; thus, it says the action and act of drawing—as well as the result, the drawing. This neglect is the basis of
the thought of form as representation, which strives to shape what
cannot be shaped, namely, the shaping, the "amorphous substance"
of forming. This is also what renders Clarice's writing of gerundive
time a "rebellion against the phenomenology of inner time consciousness."[49] Any such phenomenology remains entirely bound to find an
adequate form for the thought of forming. The thought of the form of
forming is, according to her, the way a living life—a drawing drawing
the drawing—is humanized and as such devitalized, constituting the
basis upon which the system of hope that defines humanization and
its corresponding devitalization and denaturing relies. It is by cutting
meat into pieces, cutting the is-being, gerundive time into pieces,
that life is humanized, that form is conceived. Though we must also
note that the vision of the amorphous is-being, which she also calls
the "infinite monstrous meat," of passion, is horrifying. Clarice says:

> And that this is my struggle against that disintegration:
> trying now to give it a form? A form shapes the chaos, a
> form gives construction to the amorphous substance—the
> vision of an infinite piece of meat is the vision of the
> mad, but if one cuts that meat into pieces and parcel

them out over days and over hungers—then it would be
no longer perdition and madness: it would once again be
humanized life.[50]

To approach the is-being, to eat the jelly oozing from the roach
of the gerund of being, one, the writer-reader, must dehumanize
herself, must lose the "third leg" of humanity. The "third leg," an
expression that can also be read in Sartre's discussions about engaged
literature,[51] is what renders walking impossible, turning one into a
stable tripod. This third leg is also called "a truth." To lose the third
leg of the systems of humanization, of the thought of form—this is
what is most terrifying. In *The Passion*, Clarice describes the system of
humanization using a neologism in Portuguese, *sentimentação*, which
receives the misguiding translation "sentimentalized." In the expression
sentimentação, the words "sentiment" and "sediment" are fused, indi-
cating a system of sentiments and feelings that sediment, and which
is then called a thought. In the absence of a better word, it could be
rendered to English with another neologism—"sentimentation." For
the question is not simply one of continuing or intensifying the long
civilizational fight between thought and sentiments, between reason
and feelings, or even the combat between language and presence, as
in Blanchot. It is about the courage to give up the systems of hope,
qua "sentimentation," implemented in habits of thought, feeling and
interpretation, based on these oppositions.

Clarice did not write theoretical texts and declared herself to be
entirely unsuited to critical discourses and theories on literature. She
even considered that the term "literature" was nothing but an expression
used by critics to name what writers do. She gave however one lecture
on literature at the University of Texas in Austin, about "Avant-garde
Literature in Brazil" during a conference on this topic at the International
Institute of Ibero-American Literature in 1963, the same year she finished
her manuscript of *The Passion*.[52] In this lecture, she is explicit about the
inadequacy of thinking the writing as a relation between background
(*fundo*) and form (*forma*), which she considers as disagreeable as the
dualism between body and soul or matter and energy.[53] She compares
these divisions with the impossible task to divide one thread of hair,
distinctions that are, as we say in English "hair-splitting." She is also
very critical about current notions of avant-garde as formal innovation
and suggests that, only for the sake of such a theoretical discussion, it

would be better to speak of "theme" rather than ground or content and form. She thinks of a theme as indivisible unity of background (*fundo*) and form (*forma*), and claims that this indivisible unity is that of the reading, the seeing, the listening, the experimenting. "I proposed myself: theme, and the written thing; theme, and the painted thing; theme, and music, theme, and living."[54] According to Clarice, the shortcoming of the thought of forms and its untiring divisions, such as between matter and form, background and form, content and form, form and forming, results from neglecting how the so called "formal innovations" when they are really innovations, arise more from the discovery of being free than from liberating someone or something from chains. This discovery is for her a big "creative violation."[55] Thus, to "discover is to invent, to see is to invent."[56] When a language is transformed it is because it is transformative to the extent that it demands a reading for being rather for merely reading and edification, thus, "in its apparent strangeness, we recognize that it touches our utmost intimacy."[57] Far from an aesthetic program, Clarice relates this vision to the experience of writing in the Portuguese of Brazil and calls avant-garde in Brazil the action of " 'thinking' our language," not to think about the language but to think *our* language, a "language that has not been yet profoundly worked by thought," in the sense of "thinking sociologically, psychologically, philosophically, linguistically, about ourselves." The result of such a thought can only be, she claims, "what is usually called literary language, that is, the language that reflects and says with words that instantaneously allude to things we live through."[58] With these words, Clarice also gives a hint to why Brazilian thought happens rather in the language of literature than in the language of philosophy. But if these remarks could be claimed by every cultural experience at the level of its creation, Clarice observes the specificity of the Portuguese language in Brazil, which is the one of being a language that "still boils" as she says, that appears to itself as languaging, gerundive in its own way of being as language, a "language that needs more the present than tradition."[59] It is "marvelously difficult to write in a language that still boils," that is still becoming language; what is at stake is thus the language of a life that emerges before itself as the whatever will be happening. Clarice assumes the Portuguese language in Brazil as the language of the gerund of being, and the marvelous difficulty she deals with is the challenge to note down and fix what can never be fixed: the is-being.

The Risk of Writing in Gerundive Time

Cixous recognizes herself in the writing of Clarice insofar as Clarice writes, as she says, near "the very drive to write."[60] Clarice writes the "coming to write."[61] Comparing Clarice and Blanchot, Cixous also remarks that despite the affinity of the subjects in both, in Blanchot one finds "models of tactics of avoidance," whereas "in Clarice the positions are always in movement."[62] Even if she acknowledges the strong role of the gerund in Clarice's language, she maintains the frame in which "the is that the instant is" can only be thought as a coming to be, and hence from a before being, before writing, before saying. But more than a literature of the coming to write or of the coming to be, albeit this coming may reveal itself rich of interstices and intermittences, Clarice proposes the writing of writing itself, a writing that reads itself writing, as the strongest experience of leaning the tongue on the living matter of the is-being.

How does one describe this writing in gerundive of the gerund of being, of existence, of life? It is more than a strategy of passionate writing, of "categorical affirmation" and of "denegation;"[63] Clarice prefers to call it "visual meditation," the difficult work of putting painting and music in words.

The Passion according to G. H. begins and ends with dashes, short graphic lines, ------. The opening lines are lines. These typographic lines are not a detail. They open the text by drawing the eyes to the language of drawings and traces turning them away from the language of words. The reading of lines and dashes is performed quite differently from the reading of words—it demands of the eyes a different movement, velocity, and rhythm. "What I tell you should be read quickly like when you look," writes Clarice, and even if she writes this sentence in the other book, in *Água Viva*, it gives a hint about the meaning of "gradually and painstakingly" that she demands from her readers in the *Passion*. In question is not merely the need to read slowly and let things happen and approach but to read as one *gazes* (*olha*), to look quickly, because it is only very quickly that one reads the writing being written. In this velocity, the whole reading is oriented by the risk of chance.

The risk of chance: in Portuguese, the word risk, *risco*, has a double meaning. It also means a trace, or to use an older form, a trait, a line in a drawing. The risk of chance can be understood here above all as the chance of a risk, *risco*, as the chance of the traces and traits of a drawing. A trait, a risk in this sense, is itself the risk of chance, for it has no form, no figure, being nothing but its own drawing. In Clarice, the question at stake is not about withdrawing or withdrawal, but about drawing, which English expresses clearly and without metaphors. Clarice says in different works that she aims to write as one paints, and she describes this writing-painting mostly as a drawing; thus, at stake in a drawing is how lines are drawn—indeed the fact that the drawing draws the drawing of lines. The drawing does not draw "something"; it draws the drawing. It is in this sense Clarice performs her writing-painting. That is also why it is more about "scratching than writing," as it is written in *The Passion*, indeed, to be closer to the Portuguese original, more graphite sketching than writing. We might also say that it is more about *writing down* the writing reading the writing, to insist once again in this expression. In one of her short *Chronicles for Young People*, Clarice writes the following:

> Why, how one writes? What is said? How to say? And how
> to begin? And what to do with the white sheet of paper
> confronting us calmly? I know that the answer, even if the
> most intriguing, is only one: writing . . . Beside the hours
> I am writing, I can absolutely not write.[64]

As mentioned before, Clarice begins her *Passion* (according to G. H.), which is the passion of writing the reading of the writing of the reading, drawing our attention to lines and dashes, and asking for a quick reading like when one is looking. In *Água Viva*, she explains "why the fine black lines? because of the same secret that now makes me write as if to you, writing something round and rolls up and warm, but sometimes cold as the fresh instants, the water of an ever-trembling stream."[65] The "fine black lines" are writing-drawings of the "ever-trembling stream" of the instant. Writing in gerundive the gerund of being is to put drawing into words, a difficult task. She says that "the silent word (can be) suggested by a musical sound."[66] It is not, however, a question of expressing views or impressions, experiences or experiments but to write-drawing as she listens to music, namely, "—I gently rest my hand on the record player and my hand vibrates, sending waves through my whole body [. . .] and the world trembles inside my hands."[67] Here again a dash begins the explanation of how she listens to music. It is a dash that differs from dashes in philosophical discourses such as the one that can be found at the end of Hegel's *Phenomenology of the Spirit* or at the beginning of his *Science of Logic*, which is the dialectical dash that "both holds back and propels," that "interrupts and prolongs."[68] Clarice's dash is rather an "electronic drawing without past or future: it is simply now."[69] Here dialectics, either positive or negative, either transcendental or speculative, are unintentionally left behind insofar as the issue here is "to write to you with my whole body, loosing an arrow that will sink into the tender and neuralgic center of the word."[70] Clarice's dash is showing the visual energy of writing, in which the gerund mode of writing emerges before the eyes, demanding a quick reading, that is, looking, thus her books are "pure present," a "straight line in space."[71] The awareness of how her writing is painting, understood as "graphite sketching" and "electronic drawings," writing with the whole body, and as such a writing that is necessarily a "writing-to-you," an expression she uses as a new verb for writing, is not due to a pictorial aesthetics that praises the abstract before the figurative, or to any modernist experiment with language. Reading her own rendering of what she is doing—for example when she says, "I paint painting. And more than anything else, I write you hard writing"—it is important to observe the double meaning of these phrases. They say: I paint (the) painting and I paint painting, I write (the) writing and I write writing. She is searching for writing the *is* that the instant is as a drawing draws lines. That is also why she insists on describing her writing-(electronic)-drawing as drawings and paintings entirely free from the dependence of the figure, as nonfigurative, as the painting of a "whatever it will be happening." This writing you hard writing, this writing of whatever it will be happening, the *is*, the gerund of being must be read differently; it can only be read for being.

After the short series of dashes which opens the book, the text reads:

> - - - - I am searching, I am searching. I am trying to understand. Trying to give what I've lived to somebody else and I don't know to whom, but I don't want to keep what I lived. I don't know what do to with what I've lived. I am afraid of the profound disorder. I don't trust what happened to me.[72]

These lines could be said by anyone trying to read this book. Because the reading of this book is the very experience of "trying to give what I've lived to somebody else," the experience of trying to understand, the experience of not trusting what happened to us as a reader, because the reader is afraid of such profound disorder. The writing of Clarice is not the writing of the disaster as in Blanchot, but the writing of such a profound disorder and disorganization that not only does no one know how to live in it, but no one knows how to live *through* it—that is, no one knows how to read and understand it, how to write and talk about it. But not only this: after beginning to read it, no one knows how to live without it, without searching and searching how to read it. Indeed, this book provokes a tremendous disorganization of systems of reading and of interpretation because what happens in the book also happens at the same time in the reading, as if the book were writing down its own reading. That is why every attempt to write about this book, *The Passion according to G. H.*, remains a failure, forced to face the impossibility of summary or analysis, of the reabsorption or synthesis of this writing; in each attempt to analyze, interpret, or synthesize, one wants more and more to quote each line of the book. Thus, what could be said about the book that the book itself has not already said much *much* better, more clearly, beautifully, and sublimely? It is an unequal and unfair struggle because Clarice has always already won. Maybe we should indeed do only this—read it aloud, quoting it again and again, reading it as one sings a recitative in an oratorio, rather than a prayer. The book shows the reader, already from the start, that what happens in the book happens in the reading. In order to follow this, one has to approach this instant in which the writing writes down the reading in its approaching the reading of the writing; one needs to approach this enigmatic instant, which is deeply visual.

What happens in this reading?

> Maybe what happened to me was an understanding as complete as ignorance, and from it I shall emerge as untouched and innocent as before. No understanding of mine will ever reach that knowledge, since living is the only height within my grasp—I am only on the level of life. Except now, now I know a secret. Which I am already forgetting, ah I feel that I am already forgetting . . .
>
> To learn it again, I would now have to re-die. [. . .][73]

In this book, each word, each phrase, appears as a verse, that is, complete in itself, and this to such a degree of intensity that either we read and re-read the same phrase because we can no longer exist without it, or we forget it immediately when reading the next one because it is even stronger, even more sublimely beautiful. As Cixous expressed well, "each sentence opens onto another wonder."[74] Sublime beauty means here nothing but the force of an appearing appearing in its happening. Each phrase—indeed each word—in Clarice's writing is such that we would like to keep listening to each one without rest, but, that being impossible, we begin to forget it for the sake of reading and listening to the next. Because the sense of this book, of its reading, becomes sensible only *while* reading it, we are reading the reading of this book which is the writing of its own reading—we tend to forget immediately what we have read. It is a text impossible to remember, much less to learn by heart, even while it demands nothing more than to be read by heart, through the heart, in the heart of the is-being. Clarice knows it very well, and says, "I know that after you read me it's hard to reproduce my song by ear, it's not possible to sing it without having learned it by heart. And how can you learn something by heart if it has no story?"[75] Having no stories, the book can begin wherever and whenever; it is on-going. It is a very hard reading, thus coming to the next sentence, we also cannot leave it behind, and so on. Her writing breaks the tonal experience of melody, by which Husserl tried to describe the flux of the conscious of internal time.[76] In Clarice, time is atonal, as she insisted herself. Each word and phrase are at the same time deeply connected to every other, like the lizard that remains trembling after each cut and a whole world in which one could exist for ever. She speaks about the atonal in the sense of being in the mystery of each tone setting many other tones to vibrate at the same instant. And even if Clarice says in different passages of her work that she is "fragmentary," her sense of the fragment differs profoundly from both the romantic and the Blanchotian visions of the fragment. Clarice's fragments are in certain aspects more like pigments, in others, more like vibrating tones, closer to the resplendence of energy of pearls, for moments kept within a shell. That is why the reading experiences of each phrase are something very close to the experience of an enigma. It is not so that she aims to bring things back to their value of enigma, but each word and phrase is already inside the enigma of the is-being. And what the experience of an enigma learns, says Clarice, is that "the explanation of an enigma is the repetition of the enigma."[77] We could say that her writing is such that each phrase, each word, effaces the former, so that the reading is continuously forgetting, not in order to efface or to undo language as Blanchot wanted, but to "translate the unknown into a language one doesn't speak,"[78] "to make the phonetic transcription"[79] of the is-being, of its "telegraph signals," to use some other of her frequent electric images.[80] As we saw above, Clarice describes her writing as graphite sketching, scratching, drawing, and painting, and also as phonetic transcription, as translation into an unknown language. She also describes this unknown language as the language of the sleepwalker, the language of sleep, which is the formless language of the formless. She describes it as the writing upon the wall performed by light itself, in the literal meaning of a photo-graphy, graphing figures on the wall, hieroglyphs. "The drawing wasn't a decoration: it was a writing."[81] The writing is performed as reproduction, and this in turn as the absolute opposite of expression: "More like scratching [graphite sketching] than writing, since I'm attempting a reproduction more than an expression."[82] She characterizes her writing as a kind of stenography, a fast writing that aims at noting down and hence at reproducing rather than expressing the is-being. This writing the reading of the being written presents a certain poetics of improvisation, which we see performed in *The Passion according to G. H.* and treated quite thematically in her *Água Viva*.

For Clarice, this reproduction that notes down the gerund of being, almost as an imprint, is what takes place in improvisation. Improvisation is a concept captured by musical experience. It is mostly understood as lack of planning and as formal freedom. Common discourses on improvisation affirm not so much freedom of form as much as forms that are formed without project; the very term "improvisation," from the Latin *improviso*, meaning unforeseen and not prepared beforehand, endorses such an affirmation. Improvisation in Clarice, however, is connected to the formlessness that emerges in the writing of the reading of the being written, the formless of the is-being. Because to write is to read, and to read is to see—indeed is seeing in action—improvisation is connected with the act of seeing, with the invention that, for Clarice, "to discover" and "to see" *mean*. In *Água Viva*, we read,

> When you see, the act of seeing has no form—what you
> see sometimes has form and sometimes doesn't. The act
> of seeing is ineffable. And sometimes what is seen is also
> ineffable. And that's how it is with a certain kind of think-
> ing-feeling that I'll call "freedom," just to give it a name.
> Real freedom—as an act of perception—has no form. And
> as the true thought thinks to itself, this kind of thought
> reaches its objective in the very act of thinking.[83]

Whereas most theories of improvisation hold that improvisation is
the contrary of thought, Clarice interrupts this idea, underlining,
on the one hand, that it is the act of seeing that has no form and
that the act of perception is "real freedom," and, on the other, that
"freedom" means thinking-feeling (*pensar-sentir*), an expression that
she probably heard in Guimarães Rosa, another great Brazilian author.
Thinking-feeling in one, or rather tied together in their tension,
means freedom as an act of seeing and of perception; thus, to think
here is grasped in terms of the *act* of thinking and not in terms of
thoughts. Thus, in thinking-seeing, or thinking-perceiving, one sees
and perceives the act of seeing and perceiving, and this is already
an act of thinking, "real freedom." To write this hard writing is to
write the act of freedom that thinking-seeing, that thinking-perceiving
is. What is one doing, then, in this writing the writing? *Água Viva*,
answers this question saying: "I know what I am doing here: I'm
improvising, But what's wrong with that? Improvising as in jazz they
improvise music, jazz in fury, improvising in front of the crowd."[84]
Theories and practices of improvisation are accustomed to thinking
of it as an art of variation that does not follow rules except the rule
of following and relating to what has been played right previously
without scores or plans. It demands memory and oblivion at once.
Clarice presents a different experience. She hears what "this jazz
that is improvisation says."[85] She hears the sizzling and flaring sound
of the word "j-a-z-z" the "zz" sizzling of the instant in which "meat
is devoured by the sharp hook of an eagle that interrupts its blind
flight."[86] She hears more the flash of the instant, the is-sounding-now
of the word. Her improvised writing is further described as writing
in signs that "are more a gesture than a voice"[87] and in which she
throws herself in the line of her drawing for "this is an exercise in
life without planning."[88] Writing improvisation means to obey the

order of breath and let oneself happen.[89] In this improvisation, the question "where am I going?" becomes totally meaningless, and the answer can only be, "I'm going."[90] Writing as improvising is not named, defined, or experienced as a ruleless variation of different motives, patterns, or structures. It is about the writing of the now, indeed the writing of the writing that is at once also the reading of it. It is a writing that holds its instrument—the keys in a typewriter or a pen—as one "holding a little bird in the half-closed cup of your hand," which "is terrible, like having the trembling instants inside your hand."[91] Clarice's depiction of the improvising writing is tremendously hard, quite far from the romanticism of freedom in improvisation. The whole passage reads as follows:

> The frightened little bird chaotically beats thousands of wings and suddenly you have in your half-closed hand the thin wings struggling and suddenly you can't bear it and quickly open your hand to free the light prisoner. Or you hand it quickly back to its owner so that he can give it the relatively greater freedom of the cage. Birds—I want them in the trees or flying far from my hands. I may one day grow intimate with them and take pleasure in their lightweight presence of an instant. "Take pleasure in their lightweight presence" gives me the feeling of having written a complete sentence because it says exactly what it is: the levitation of the birds.[92]

These lines sum up both Clarice's vision of improvisation and the way she sees it while improvising, that is, writing. Improvisation is thus not the "whatsoever may be written and played," but rather the mad lucidity of complete attention to the is-being, to the is that the instant is, *in* the writing of it, which is the attempt to hold it in the half-closed hand of the one who is typing, who is writing or drawing. At the core of it is the struggle between holding what cannot be held and the unbearability of holding it any longer. She speaks here of the attention to the sentence being written *at the moment* it is written, the sentence that writes—and hence tries to hold—the passing-through of the is being written now, this "lightweight presence," "the levitation of the birds." This improvisation is about "writing to you in time with my breath."

It Is-Being: Or the Neuter Crafting of Life

The literature of the is-being, of *sendo*, to recall the Portuguese word, is a literature of nothing. Nothing is happening except the is-happening, its unfolding. The is-being cannot be taken as an event, understood in the sense of an interruption of a continuity. It gives itself as precisely the opposite, as the continuation of an interruption. Indeed, this is what the cockroach of the is-being exposes—the continuation of an interruption, a lizard that continues to tremble and squirm, after being cut in pieces, either by life itself, or by what is said and thought of life.

Clarice rephrases or, better, rewrites the discussions and searches for transcending transcending, overcoming overcoming, and thereby for finding a way out in no longer trying to force a way out. Heidegger and Blanchot come close to a thought of gerundive time, which is a thought of the is-being, without being able "to lean their mouth on the matter of life." As Cixous put it, "Heidegger wrote on the pitcher but not on the chick,"[93] which is almost a totemic animal for Clarice. They both discuss it in terms of the neuter—*Es, il, Es gibt, Es ist, Es gilt, il y a*. Clarice does as well. She, however, does not only speak or think of the neuter; she rather speaks and writes in neuter. The writing in neuter shows the impossible grammar of "life being seen by life," of "the great neutral reality" affirming that "the most important word of the language has one sole letter—*é*, which means "is" in Portuguese[94]—saying the impossible neuter using the neuter *it* in English as she does for instance in *Água Viva* (the Portuguese language does not have a neuter pronoun) or the X, and further employing the expression "the neuter" as a pronoun and as the subject, the only word one can interchange with the cockroach, the cockroach of the neutral ancestral is-being. "Now whatever is luring me and calling me is the neuter. I have no words to express, I speak therefore in the neuter. I only have that ecstasy, which also is no longer what we call ecstasy, since it is not a peak. But that ecstasy without a peak expresses the neuter of which I speak."[95] The language of the saying and writing in neuter demands impossible sayings and writings. It no longer conjugates "to be" but the very "is" as a reflexive verb, as in "to is itself," "to is yourself."[96] In Clarice, the language of the neuter is a language in which the verb "to is" becomes not only possible but also inevitable. Thereby she turns language into an ancestral murmur.

Unlike the insistence on the neither-nor, Clarice's neuter names the things' *is*, the "it" or X, in which the abstract does not really exist. The neuter is "the figurative of the unnamable"[97] is-being. This is a strange figureless figurativeness which can be understood as a "photogenicity," given how this language in neuter of the is-being can "catch and register the invisible light irradiated by objects and, at the same time, to use one's sight in order to call things forward into visibility from their depth and darkness."[98] It is a way to capture the "visual meditation" performed in this writing that is continuously reading its being-written, and in which things are neither called into presence nor acknowledged in their opacity for only being able to be seen through the lenses of language. The improvising writing in the neuter, in the gerund of being, the writing that conjugates the verb "to is," shows how we—that is, language—are as things are, namely an "is," things looking at other things, an "it" as much as every other "it," that the is is. Thus, as much as all other things, words are nerves, breaths, and electric signals, radiating respiration and radiance. This does not mean that language and things, that the human and the animal or every other form of life are the same; the drama of this difference cannot be erased. This drama appears, however, to be even more dramatic when one discovers that things are their *is-being* as much as language is nothing but the is-being. This is experienced precisely in such a writing as it is struggling to hold the "is" and being defeated by it, which, being much stronger, forces the writing to open its half-closed hands to let it fly again; thus, the "instant-now is a firefly that sparks and goes out, sparks and goes out."[99] What appears here is "the world: a tangle of bristling telephone wires"[100] (in Portuguese, "o mundo eriçado de antenas"), millions of things sending telegraphic signals of the *is* that does not belong to them but to which they belong.

In his writings about the neuter and literature, Roland Barthes insists that literature is a thought of the neuter insofar as it thwarts paradigms [*déjoue les paradigmes*].[101] It thwarts paradigms insofar as the neuter is a "va-et-vient," an "amoral oscillation," the contrary of an antinomy. A writing in neuter such as Clarice's not only thwarts paradigms but also the paradigm of the paradigm itself, the very form of form. The thwarting of the form of form provides another way to formulate her search of writing down the formlessness of the is-being. If we consider the literary-theoretical treatment of the neuter in Blanchot, for whom the neuter should transcend the thought of being, the neuter in Clarice proposes another direction of thought. It proposes a path for thought that leads away from the problem of deciding between being and not-being and towards an attempt to listen to the murmur of the is-being of being: *o sendo do ser*. The is-being of being is embraced by Clarice's writing as the ancestrality of being, its immemorial time that exposes the continuation of an originary interruption, the interruption of life in and by each life, each life being the hacked-up part of the neuter lizard of life. This is neither tradition nor the event, not even existence *as* an event, because what is happening is *nothing*, is the "nothing-happening."[102] At stake is not really an event but "the neutral crafting of life,"[103] "the neutral present of life,"[104] "the neutral life that lives and moves,"[105] the neutral love, the neutral God, the neutral plankton,[106] my living neutrality,[107] "the great neutral reality."[108] "The neuter. I am speaking of the vital element that binds things,"[109] she insists. Clarice's writing differs greatly from that of Heidegger and Blanchot, then, in that she does not identify the neuter with the thought of the event, even when the event is understood as overcoming overcoming. Some readers of Clarice want to attribute to her a thought of the event, recognizing either a Spinozist version that considers the event as continuous or discovering in her an event qua epiphany, due to its uniqueness.[110] Considering that in both Heidegger and Blanchot the thought of the event is indebted to a thought of truth as unconcealment, as appearing while withdrawing, we find in Clarice a quite different truth, the truth that "the truth is what it is,"[111] referencing the way she rephrases the biblical text. The neuter says what can only be written and said in neuter, namely, "living life instead of living one's own life—entering divine matter, losing individual life, losing the organization—the sentimentation of the human world."[112] This writing *in* neuter—not only about and of the neuter—writes down "existence existing me,"[113] taking the verb "to exist" as transitive in order to express how one being is "existing" the other being,[114] how "I was me being," *eu estava me sendo*,[115] and how life is me, *a vida se me é*,[116] and "its [the crockroach's] existence existed me [*a existência dela me existia*]."[117]

Writing in neuter is writing in gerundive time. This writing in the gerundive time of the neuter should be analyzed more carefully through a gradual and painstaking treatment of Clarice's own creative use of Portuguese grammar, her verbal tenses and forms, the rich sonority of her phrases and words, in which the gerundive character of the Brazilian Portuguese language becomes very clear—such a task, unfortunately, does not fall within the scope of the present book.[118] Gerundive time is the tense in and upon which Clarice's writing lives, and from which the gerundive emerges as "this verb on a horseback," to recall the poetic image used by Osip Mandelstam.[119]

Rather than a temporality of repetition, the gerund sounds like echoing resonance. We could speak here about echography. *The Passion* is written as an echo-graphy, a writing down of echoes, rather than as a kind of literary art of the fugue. The book is structured so that each chapter begins with the last phrase of the previous chapter. This repetition hints at the meaning of this echo-graphy. It should not be read as if Clarice were imitating a musical fugue but as the search to writing down the listening to the echo of how language as such sounds in the throat, coming directly from nature, letting the sounds of earth reverberate in human and animal throats. In Clarice, we should not search for the opposition between language and presence, which is the Blanchotian version of the old metaphysical distinction between language and nature, for language—if it is in a certain sense the human—is in Clarice's writing an "orgasm of nature."[120] In this sense, Clarice is closer to Schelling, an author that remained unknown to her. Her first novel, *Near to the Wild Heart*, begins with noises: "Her father's typewriter went clack-clack . . . clack-clack-clack . . . the clock awoke in dustless tin-dlen, the silence dragged out zzzzzz. What did the wardrobe say? clothes-clothes-clothes."[121] It is indeed in this first novel to express a piercing understanding of the relation between language and nature insofar as language is understood first of all as *voice*, and the voice of language as what is of the earth, voice that, without smashing any object, arrives at one's throat smoothly and from far away as if it had crossed long paths under the soil.[122] Language is this enigmatic crossing of long paths under the soil achieved by the voice of earth; language appears on earth as the echo of the voice of earth and each language as the echo of the voice of language as the echo of the voice of earth. There is no opposition. For Clarice, language articulated in words and names is a way to search for "the great neutral reality," a way to search for it and without finding it, but thereby learning precisely to listen to the murmur of the voice of language echoing the voice of nature in every throat, in every language.

Language is a way of searching and not finding. It is not a detour but, quite to the contrary, the only way, the only tour and the only chance, she insists in her writings. Searching—that is, in language—one comes back *with* empty hands, one comes back *with* the without. One comes back with the enigma, with the unsayable and unnamable is-being, reality, in her arms, her poor human arms. As Clarice wrote:

> Reality precedes the voice that seeks it, but as the earth precedes the tree, but as the world precedes man, but as the sea precedes the vision of the sea, life precedes love, the matter of the body precedes the body, and in turn language one day will have precede the possession of silence.
>
> I have to the extent I designate—and this is the splendor of having a language. But I have much more to the extent I cannot designate. Reality is the raw material, language is the way I go in search of it—and the way I do not find it. But it is from searching and not finding that what I did not know was born, and which I instantly recognize. Language is my human effort. My destiny is to search and my destiny is to return empty-handed. But—I return with the unsayable. The unsayable can only be given to me through the failure of my language. Only when the construction fails, can I obtain what it could not achieve.[123]

If in a first reading of these lines, the civilizational opposition between language and reality seems acknowledged, it is important to remain attentive to the "but" she inserts right in the first sentence. Reality precedes voice "but" as the earth precedes the tree, meaning how life begins in life, how language arises within reality, within the is-being and not anywhere beyond it. Clarice writes—*in* gerundive time—the gerundive time of the right now, which demands the abandonment of all systems of hope, of all forms of humanization and sentimention, that is, of thought, feeling, and form. She thus performs what she once called in conversation with her son the "ex-possible," saying and thinking of is-being *while* is-being, of reading the writing *while* writing, of writing the reading *while* reading. This writing of the is-being is the writing of the one that discovers how life and existence are nothing but the *is*. This is the most profound experience of existence in exile, a tremendous experience, of, in the midst of the utmost confinement, getting the chance to become un-delimited and

> to have the courage to use an unprotected heart and keep talking to the nothing and to no one? As a child thinks [about] for the nothing. And run the risk of being crushed by chance.[124]

> To keep talking to the nothing and to no one, to write to the nothing and to no one, to write-to-you—this is supremely frightening, like a child wandering the earth alone—so needy that only the love of the entire universe for me could console and overwhelm me.[125]

In these phrases, we find an eloquent (and somewhat biographical) figure of exile in Clarice, the figure of a child wandering the earth alone or of a blind person roaming about in open fields. Differently from the claim that, if exile makes one fall silent, the struggle should be for an exile that makes earth and produces the opposite to silence, extinction of voice and breathlessness,[126] Clarice would like exile to make one fall in the *is* and discover, what means see "—my reachable present is my paradise lost."[127]

Clarice rarely speaks of exile and leaves behind the figures of excess and ecstasy.[128] Discovering and seeing—that is, *inventing*—the grammar of the "to is," she is the great cardiographer of exile. Thus, what remains in exile is nothing but the gerund of being.

Chapter 5

Without Conclusion

A Home in Gerundive

Which tense do you want to live in?
  —I want to live in the imperative of the future passive participle—in the "what ought to be."
  I feel like breathing that way. That's what I like. There exists such a thing as mounted, bandit-band, equestrian honor. That's why I like the fine Latin "gerundive"—that verb on horseback.

—Osip Mandelstam

In the preceding chapters, a thought of the is-being, of the gerund of being and existence was sketched out for the sake of rendering the core of the experience of time from within exile. It has been stressed repeatedly that exile is an immense struggle for presence, in which existence is entirely exposed to the unsheltered is-being. Facing the need to dwell in the unprotected and groundless is-being, existence in exile tends to flee and escape from it, either through nostalgic utopias or utopian nostalgias, or even through the alienation of discourses of a consumerist "live in the moment," or "seize the day." But how to dwell in the is-being, how to be in the is-being without fleeing from it, if the is-being is continuously fleeing from itself? How to bear the terrible task of holding this little bird of the is-being in the half-closed cup of one's own hands, to recall Clarice's image? In fact, the question of dwelling and inhabiting is the most dramatic in exile. Exile is searching for asylum, not only for a home that would end the state of homelessness, but for a home *in* the homelessness that never ends, not even when one finds an asylum. In exile homelessness is without end.

In 1934, the Russian poet Marina Tsvetaeva, one of the greatest poets of exile in modern literature, wrote a poem called *Toska po rodine*, meaning, in one of its English versions, *Homesick for the Motherland*. The poem reads as following:

> Homesick for the Motherland! Long
> Unmasked confusion!
> I do not care—
> Where I am completely lonely
> Or over what stones I wander home
> With a shopping bag
> To a house, that is no longer mine
> Like to a hospital or barracks.
> I do not care that I am among
> Bristling people—a captive
> Lion, or what human society
> Will cast me out—as it must—
> Into myself, my individual feelings.
> A Kamchatka bear without ice
> Where I do not fit (and no goodbye!)
> Where they grovel—I am one.
> I will not be seduced by the language,
> The mother tongue's milky call.
> I do not care—in what language
> I am humiliated!
> (Or by what readers, newspaper
> Swallowers, searching for gossip . . .)
> They belong to the twentieth century—
> I am—before all time!
> Stunned, like a log,
> Left over from an alley of trees.
> People are all the same to me,
> And I could be just equal to—
> A former native—only.
> All my tokens, all meanings,
> All dates—are gone:
> My soul, born—somewhere.

For my country has taken so little care of me,
That even my keenest eye
Along with all my soul, all—have been alienated!
That even my birthmark cannot be discerned!
Every house is alien to me, every church is empty,
Everything and all—is the same.
But if along the road—a bush
Rises, especially—a rowanberry . . .[1]

In this poem, we listen to words that belong to millions of mouths and millions stories of exile: the complete loneliness, houses that are no longer one's own, the being cast out by human society, the feeling of being a polar bear without ice, of not fitting into any place but also being unable to say goodbye, the experience of one's own country taking little care of its own people, of being a "former native," of having all tokens, all meanings, all dates gone, and not even the birthmark being discerned. In this experience one not only loses a home but also "every house is alien to me," "every church empty," one's own soul and all become alienated and "everything and all—is the same." The poem describes the last century as the century of homelessness and above all as the century in which home itself became alien to itself. Without any attempt to interpret this poem in all the levels of its density and poetic creation as much as in its relation to Russian poetic tradition, an important lesson about contemporary exile can be learned, namely, that contemporary history is a history marked by the fact that *home itself became homeless.*

In which sense does home itself become homeless in contemporary history? The poem insists that it does not matter where one is and in which language one is humiliated; not even the "mother tongue's milky call" can be a home any more. Not even language remains to protect the one who has become homeless. "Everything and all—is the same." Following the poem, we can say that home becomes homeless when a home has no longer a place and a language, when the tight bonds between home, the land, the place, and the language are cut and overturned. That is, when motherland—a word that binds tightly the land and the language, connecting the fatherland and the mother tongue—is no longer *at home.*

This poem, written 1934, sings the tragic song of the loss of the meaning of home in times where home became the most frightening and alienating political weapon. The age of terror and totalitarianism that marked the first half of the twentieth century—and which continues not only to haunt us, but also reemerges to menace with new force and new weight today—is the age of the lethal politics of homeland, through which home became the place of extermination, terror, and repression. If, for centuries, Europe has exterminated and destroyed other people abroad through violent strategies and techniques of colonization for the sake of building a homeland in the world, the twentieth century made the experience of the homeland into one of exterminating and destroying itself, precisely at the moment the experience of home connects to the ideas of a land, a place, and a language. In this century, home showed itself as threatening, haunting, and uncanny. In fact, home, land, place, and language—altogether became homeless, frightened, and frightening, to the extent that their bond gives rise to terrible nationalist politics of persecution, segregation, and destruction, even while their disconnection provokes the extreme sufferings of homelessness, not belonging, and exclusion. It is indeed in regard to this bond that one of the deepest aporias of the century emerges, namely, that the moment this bond is built and defended in nationalist views, it is dissolved insofar as, by its segregating violence, it results in exile, expulsion from home, land, and language.

"Uncanny" is the English translation of the German *unheimlich*, a word that negates home from within. Already 1919, Freud understood, at the psychological level of individual experience, this internal negation of the experience of a home, in his well-known essay *Das Unheimliche, The Uncanny*.[2] His thoughts on *das Unheimliche*, the uncanny, depart from the semantic web provided by the German language, in which *Heim*, home, *Heimat*, homeland, *heimsich*, native, *unheimlich*, uncanny are connected with *geheim* and *Geheimnis*, with secrecy and the secret. He develops a definition proposed by Schelling, in which "*Unheimlich* is [considered] the name for everything that ought to have remained . . . secret and hidden but has come to light."[3] The uncanny is the coming to light of what home has hidden and repressed inside itself. It reveals that home is not only the place where one is protected and secured but also the place in which one is frightened of being robbed of their own eyes, that is, frightened of repression and castration; the uncanny is the coming to light of home as a place of haunting ghosts, the ghosts of self-doubling and repeating itself, the place in which one experiences the "omnipotence of thoughts" that leaves no free space for otherness and creation.

Freud's essay brings to light how the homely security and protection keep secret—in its secrets—insecurity and exposure. Home is not only a place that some are allowed entry while others are forbidden it; it is also a place without a way out. That is why home is also horrifying, which in Swedish can be said with the expression "*hemskt*" also coming from "*hem*," home. While common sense takes for granted the meaning of home as the place where one feels safe and protected, where one can keep her secrets, and therefore have secrets without secrets, Freud shows how home is a place where secrets have secrets of their own, their own haunting ghosts and threats.

What Freud wrote about the individual experience of the uncanny of home becoming *unheimlich*, uncanny, when home negates itself from within, also showed its truth at the level of historical and common experience. The catastrophe of the twentieth century is very much the catastrophe of the uncanny experience of a home destroying the meaning of home. Since then we have to ask: "How much home does a person need?" This question titles another well-known essay, written by the Austrian Jewish author Jean Améry, in the 1960s, that can be brought in conversation with Freud's thoughts on the uncanny.[4] In this essay, Améry accounts for the meaning of being in exile from the Third Reich as the uniquely terrifying experience of a home that exterminated the experience of being at home, of a mother tongue that exterminated the experience of a mother tongue, the experience of a home dismantling the past piece by piece. Because a home is, for Améry, above all "an access to reality that consists of perception through the senses,"[5] he experienced and witnessed not only the loss of the sense of home but further the loss of home as access to the senses and to sensibility, and hence as an access to existence. According to him, what was most terrifying in this experience, however, was to have no other possibility left for human existence beyond longing for and wanting this hostile home and this inimical mother tongue, no other possibility beyond desiring and searching for the loss of home and the loss of language as the only home and language that remain. What he witnessed, however, was not only the loss of home and of mother tongue, the dismantling of the tight bond between home, place, and language, but the substitution of home for religion and money, for acknowledgment and esteem, indeed for what he called "the exchange of home for the world" that becomes even clearer after the War. This substitution or exchange—*Ersatz*, to use his own expression—is for him the mark of the "mutation of the human being,"[6] the unavoidable "psychic assimilation of the technological scientific revolution."[7] What characterizes this mutation or psychic assimilation of the technological scientific revolution, inaugurated by modernity and accomplished in the twentieth century, is the transformation of human beings into what Marcuse called the "one-dimensional man"[8] and its world, a world where everything and all is the same. It is the same in the sense that every meaning and determination loses its meaning, becomes void of signification, and thereby ambiguous, insofar as it can receive any meaning and determination whatsoever, depending on the interests that orients the uses, misuses, and abuses of everything. This is a way to describe, as we indicated in our intro-duction, the logic of "general equivalence," which governs the process of globalization or global capitalism. It is the logic of everything being exchangeable, including exchange itself. That is why nothing can remain at home and everything must be expulsed from home. Capitalism needs war and crisis; it needs destruction and expulsion because it needs continuous moving and transformation, insofar as it is through this continuous "becoming" that it remains the same. That is why, as already stressed, continuous transformation produces continuously not only new products and forms but also and above all conformism. Indeed, what is at play is not really transformation but substitution, Ersatz, and exchange, through which every meaning can be taken as substitute for another, through which everything becomes exchangeable. This means to turn not only everything but also every meaning and sense into a mere commodity. When all meanings become empty and thereby ambiguous, the need for strong meanings emerges. When all figures and configurations become void, the search for strong figures and figurations increases. Today it becomes clearer that it is the economic politics of boundary-crossing that generates the aggressive politics of segregation and exclusion that expands over the world. It is within this liquid economy of continuous exchange that new fascist movements, new right-wings positions, renewed old fundamentalisms of every kind, can emerge so powerfully as movements in search for strong figurations—figurations building typicality. They are responses to the fallen figures of everything—not only of revolution—accomplished by global capitalism, that is, to the capitalist command to leave behind all traditional, situated, localized figures and forms—and this through violent wars of invasions in all possible senses and meanings. That is why the global world, the world that transgressed all limits, frontiers, and determinations is the world of the growing anew of an aggressive politics of nationalism and ethnocentrism, of racism and discrimination. More than a coming back of the 1930s, the backwards longing, the "retrotopia" to use the term suggested by Zygmunt Bauman,[9] what we encounter today is for sure a longing, but not a longing for the 30s, nor for known figures of the past, but rather a longing for figurations of the known, indeed a longing for figurations—which are the strongest forms of the known. As the promise of having a home everywhere, globalism becomes more and more the nightmare of losing everywhere the meaning and feeling of a home. The more home is displaced to the world, the more home moves far from home.

The last century is the century in which both the uncanniness of home, of the homeland, of the homely came to light and in which home became homeless. Home negates itself when expanded within itself and out of itself. Expanding inward in nationalistic discourses, home destroys itself. Expanding outward in internationalist discourses, home becomes strange for itself. Expanding inward in discourses for which home is only ever "here," home undermines itself. It becomes uncanny, *unheimlich*. Expanding outward in discourses for which home is everywhere, home opposes itself. It becomes homeless, *heimatlos*. How to exist, then, in a world where even the uncanniness of home becomes homeless? To exist in a world where home became homeless is to exist in a world where one cannot go back home nor find a new home. It is to exist in a *tensioned between*, between the impossibility of returning home and the impossibility of arriving at home.

To exist in a *tensioned between* marks exiled existence. In exile, one is never here because one is always also there; neither can one be there because one is also here. The tensioned between that marks the situation of exile is a place that cannot be separated from time, and a time that becomes itself a place. The before is always present in the after as much as the after mixes with the before in such a way that it becomes impossible to distinguish one from the other. That is why, in the situation of exile, the boundaries between fiction and reality are blurred, and thus one is always far from their own nearness and near to their own distance. In the tensioned between of exile in which one is no longer at home in their own home and has to feel at home in what is not their own, all images are like superposed photographs, one upon the other, as much as all words are clusters of superposed meanings, sounds, and accents. One and the other, before and after, here and there, this meaning and the other, this sound and the other, go like an echo, always one after the other and nonetheless at the same time. It is a situation of neither-nor; but not only thus it is also of both one and other.

In the huge amount of literature about exile and the experience of homelessness that accompanies it, the focus has been on the cut that separates one from oneself, individuals from their tradition and communities, the now from a past, a cut that is so extreme that it usually even cuts oneself off from a future, which is to say, from a new life. In focusing on the traumatic cut, however, such discussions about exile tend to neglect what is most sensible in the experience in exile, namely, the struggle for being present to one's own present, indeed, the struggle for presence. The real struggle in exile is not so much the struggle to not lose the past or to forge a future, but to be above all present for the present, to be where one is. Indeed, what escapes the escaping that exile always implies—no matter how voluntary or involuntary exile may be—is neither the past nor the future, but the present and its disquieting and restless presence. What haunts exilic existence is the density and weight of the "is-being." It is unsurprising that the situation of exile can produce nostalgic and utopian thoughts, which are sometimes even more tyrannical than the tyrannies it is fleeing from. In the situation of exile the unanchored and groundless character of the present, its fugacity, impermanence, and transiency, becomes so exposed, nude, and crude that one can hardly bear it alone. One tries then to escape from it, searching for strong figures and figurations, to secure oneself in a stable home, a home to protect oneself against the unanchored present, the present which is the only ground one can have in exile; thus exile confiscates both the past and the future, since one will no longer be at home at home and one will always feel foreign in whatever home one might construct. The feeling of nostalgia so discussed in the research and literature of exile tends to hide what exile brings to light—the longing for and desiring of the present, which exile continuously sequesters.

The previous discussions intended to indicate that exile is a verb conjugated in present tense. Exile says—restlessly, ceaselessly—the difficult phrase: I am, not being. "I am, not being"—says the present, with different interpunctuations and breathings, insofar as the present is, on the one hand, the disquiet of not being able to remain where it is, and, on the other, the torment of not having anywhere else to go than to remain in this impossibility of remaining. The figure of a tensioned and restless between corresponds to the disquiet of the present, in which the one who exists in the situation of exile is caught and held, or, to use a legal term, is "arrested," that echoes through the whole work of Blanchot. The present is the tense and the tension where no one can rest; it is thus always far from itself, slipping away and withdrawing from itself. To be arrested in the restlessness of the present is another way to formulate the situation of exile. How to define this arrest, this hold-up in the restlessness of a here and now that can never be measured, but from whose lack of measure all measures are drawn? Many refugees, many people in exile are today in arrest, extra-legally imprisoned, confined in places of waiting and custody, arrested in places that render existence impossible. These arrests are not only places of imprisonment but very much places in which the here and now undermine the is-existing of existence, places in which the present undermines presencing. Thinking about the contemporary experience of these places of arrest where today many people "live," the question of home receives another frame—it becomes in fact the question of how to stay in the present from which no one can escape, but in which no one can remain.

In the situation where neither the past nor the future can be a home, where nostalgia and utopia have tested their power of destruction, there is no other home left than the present, the tensioned between as such. If it is possible to speak of home today, then it cannot be discussed in terms of a bond of land, place, and language. It has to be connected with the difficulty of being a presence for and in the present, with the need to dwell in and thereby to remain close to the distance to itself that the present is. Thus, the present is perhaps nothing but an insurmountable self-distance, insofar as it is always ahead of itself, beyond itself, pure fugacity and transiency, anchored in its own lack of anchorages. Indeed, the question about home has little to do with having or not a place to live—this is a question that can and must be solved through socio-economic politics and policies—but rather to do with the question of how to dwell and sojourn in what is left, which is to say, in the is-existing of existence. It has to do with the question of having to learn to dwell and sojourn in the only place one is, the place one can never get out of it, despite all attempts to escape from it, which is the presencing of the present, the gerund of being and existence.

It is striking that the philosopher who insisted throughout his work on weaving together home, land, place, and language, the reactionary philosopher of the Black Forest and black notebooks, was the philosopher that most clearly called our attention to the fact that home is not a noun but a verb—namely, Heidegger and his thoughts on home as dwelling and more precisely as "learning to dwell." Heidegger's thoughts on *Heimat,* on homeland and on language as the home of being are far more complicated than fascist discourses on home and homeland.[10] His thoughts on home and homeland arise from his view on the homelessness of being and on how the homelessness of being has built the philosophical civilization of the West. For him, Western history is the history of a civilization without a home for being, a civilization expelling the experience of being from home. In question is not so much the lack of the experience of home in the modern world, but the lack of the experience of being as the only home of existence. In question for him is a civilization that substituted generalizations and typologizations for step-by-step experience, and thereby lost out on the way the language of experience, having substituted and exchanged it for the language of reification, calculation, instrumentalization, and control of experience. The question that occupied Heidegger for most of his life was, How does dwell in the homelessness of being, in a world where the experience of being has no home? At the end of the essay entitled "Building Dwelling Thinking," written 1951, Heidegger says that "the real dwelling plight lies in this, that mortals ever search anew for the nature of dwelling, that they must ever learn to dwell."[11] The German word for dwelling is *wohnen,* from which *Gewohnheit,* habits come from. In order to keep in mind this connection, we could say in English inhabit instead of dwelling. The Heideggerian lesson could be then summed up as: *To inhabit means to have to learn to inhabit.*

But even if we would agree with Heidegger and admit that home means to learn to inhabit, there is still something to be learned. Still to be learned is how to learn. What history—and especially the last century—seems to demand, however, is more than to learn. It demands of us, perhaps more than ever, "to learn to unlearn," to borrow a verse by Fernando Pessoa, the meaning of home that has been guiding human existence on earth for millennia. Learning to unlearn is a hard pedagogy; it demands the capacity to exist without fixed meanings, without searching for even stronger meanings and figures or confounding the abandonment of meanings with the nihilistic position that considers that if nothing has a stable and established meaning then everything is allowed. The question becomes then, How does one exist in the figurelessness without searching for totalizing figures? In this case, to learn to unlearn habitual meanings of home might be to turn the eyes, ears, and hands of attention from the meaning of home as a safeguard against the world and let home leave home. That is what Tsvetaeva noted down with poetic clarity in one verse of her *Poem of the End*, which says: "Home means: out of the house/ And into the night."[12] Learning to unlearn habits of inhabiting the world through the construction of walls that separate in and out, self and other, homely and foreign, means above all to learn to exist without, or to be more precise, to *learn to exist with a without*. This is perhaps the deepest lesson of exiled existence. Thus, to exist in exile is to exist not only without the comfort of a home that thinks, says, and experiences *for* you, *in spite of* you, *instead of* you, but to exist with the without, which demands that one thinks and says, that is, *dwells in* existence on the basis of the experience of existing arrested in the restless present, in the is-being, or to say it with the thinking-feeling simplicity of Clarice, in the "is" of each thing. This is quite different from taking the present as the mere here and now, which should be used up as quickly as possible, this way of avoiding the present by way of imputing to it the meaning of a void "now." The ideology of presentism, of *carpe diem*, of a "let us live the day" for it will soon disappear, denies to the present its dense presence, covering its tensioned and restless between with the promise of a quick flight from it.

This lesson in gerundive temporality unfolded through the experience of exile is a lesson in dwelling in the gerundive, which is to say a lesson in existing without strong figures and figurations, a lesson on how "line becomes existence" when, instead of asking "Where am I going?," one answers with precision: "I'm going."[13] It is a lesson in existing as a drawing drawing lines being drawn, which is "the figurative of the unnameable," not in the sense that it is unnameable because it is beyond the name, but rather because it is precisely the nam*ing*, the seeing, the saying, the thinking while naming, while seeing, while saying, while thinking. This lesson is a lesson in the chance of a real freedom, which, continuing to follow Clarice, is the one of thinking-feeling this whileness, the whileness of the is that the instant is. It is all about the gift of attention to each thing's *is* that the experience of exile renders possible in the midst of the impossible. In a passage from *Água Viva*, Clarice asks explicitly for it: "Pay attention and as a favor: I'm inviting you to move to a new kingdom."[14]

It is also with a sign to pay attention to each thing that Tsvetaeva ends the poem that opened this chapter. In the last strophe we encounter the verse:

> Every house is alien to me, every church is empty,
> Everything and all—is the same.
> But if along the road—a bush
> Rises, especially—a rowanberry . . .[15]

These lines indicate the need not to substitute home with anything but to turn the eyes, ears, and hands of attention to the language of the singular—the language of each thing. It demands the attention capable of unlearning the habitual confusion of singularity with identity. In question is the need to find a language of nearness, of closeness to the insurmountable self-distance that defines existing while existing, which in turn defines the is-being. If it is possible to find several affinities between Tsvetaeva's and Clarice's poetical force,[16] a decisive difference between them is nevertheless to be underlined. When Tsvetaeva calls for attention to a bush rising, especially—a rowanberry, she calls for it by way of the conjunction "but" that links through contrast and opposition. In Tsvetaeva, the attention to the singular and contingent is named as a resistance to the loss of every home in the world, indeed to the loss of home in every home. This loss that results from the violence of totalitarianism bears witness to the no-way out of a totality that expands infinitely, so that even what opposes to it becomes totalized by it. Having known the horror of totality expanding itself infinitely, that is, totalitarianism, Tsvetaeva proposes nonetheless the attention to a rowanberry that could be possibly be encountered on the road as a "resistant rest" and a "strategy of contingency," following some expressions suggested by the art historian Jean-Marie Pontévia,[17] to the infinite totalization of the world. As such the singular appears as "rest" and "remainder," of a loss that in the impossibility to fill the lacuna delivers the lacuna back to its meaning of a lake or lagoon, that is, of a void without void from which one can live and exist.

Another strategy of contingency in which a rest resists to the loss of the world in a world exhausted of the world, is recognizable in some lines in the poem *Ash-Wednesday* by T. S. Eliot that read:

> If the lost word is lost, if the spent word is spent
> If the unheard, unspoken,
> Word is unspoken, unheard;
> Still is the unspoken word, the Word unheard,
> The Word without a word, the Word within
> The world and for the world.[18]

It is the strategy of listening to how the unspoken and unheard word *still* is unspoken and unheard. This still-being gives an indication of how the exhaustion of still-being gives birth to a still-listening-to where from and where to things are as they are. T. S. Eliot calls for hearing—in the same—another tone and accent, and thereby not searching for a beyond what it is. It is a resistance through a certain sense of redemption and restoration "with a new verse the ancient rhyme," a sense that belongs to the atmosphere of this long poem, written 1927 right after his conversion into Anglicanism. Nonetheless, this sense belongs to a poetics of loss, of the absence and withdrawal of meaning that should be restored through the very loss, a redemption accomplished by the very condemnation.

The narrative of such a loss and its mourning rings quite differently from the language of Clarice. As she says: "[I]n writing I try to see strictly in the moment in which I see—and not to see through the memory of having seen in a past instant."[19] Clarice doesn't speak about the loss of home, which is always the loss of a world. She speaks from within the is-being, which is neither a loss nor a not-loss. She speaks, which means she writes from within a wild nearness to the is-being that is so narrow and tight as "the wheel of the speeding car (that) just barely touches the ground. And the part of the wheel that still hasn't touched, will touch in that immediacy that absorbs the present instant and turns it into the past."[20] To dwell the nearness of the is-being is to step, with the rapt attention of a tightrope walker, narrow and tight paths, and as such to let oneself be conducted by this narrow nearness. This does not mean to forget "history" and the works of memory, for she knows very well that "you can't walk naked either in body or in spirit."[21] What she writes is another sense of history and memory that emerges from the attention to the things' *is*, to the each is, in which each thing's is is seen "gradually and painstakingly," as a hand seeing the drawing been drawn, the writing reading the being written, a thinking-feeling conducted by the gerund of being and existence. She insists, "My story is living."[22]

Also claiming attention to the minutiae of each thing, each breath, each gesture, indeed to the "acute accent" of the today, Paul Celan used the expression *Engführung*, which can be heard as a description of this conduction by the narrowness of the gerund of being, an expression that translates into German the musical term *Stretto*, meaning in a musical fugue the imitation of a subject in a close succession. Corresponding to the Greek *agcho* and the Latin *ango* from which *Angst*, Anxiety is derived, the German "*Eng*" means the enigma of tight and narrow nearness. The Greek language has the term *engus*, nearness, in opposition to *tele*, distance. Listening to this vocabulary of tightness and narrowness of the experience of existing too near to the is-existing of existence, it would be possible to speak here of the *engnigmatic* is-being, the enigma of a home in gerundive. If we consider together with Celan that exile is the experience of language, of word, of home in the risk, what in English can be said with the expression—"in the balance" which renders well the German *Waage* echoing the vague, as he writes in the verses *Sprachwaage, Wortwaage, Heimat/Waage Exil*, it can be understood as the difficult experience of the *engnigmatic* path of nearness. This word indicates that the language of nearness is the one of a "writing to you the hard writing," a writing that, for Clarice, means the real freedom of thinking-feeling how "being (is) existing the other being" and how "the wings of things" are still open.

Notes

Introduction

1. Seneca, *Consolation à Helvia. Dialogues* VI, 1 (Paris: ed. Belles Lettres 1975), "Nullum inveniri exsilium potest; nihil enim quod intra mundum es alienum homini est." 64.

2. Plutarch, *Moralia*, vol. VII, Loeb Classical Library edition (London: Heinemann, 1959).

3. Schelling, J. W. *System des transzendentalen Idealismus* (Hamburg: Felix Meiner, 1957), 297, translated by Peter Heath, *System of Transcendental Idealism* (Charlottesville: University Press of Virginia, 1978).

4. Werner Jeager. *Paideia: Die Formung des griechischen Menschen.* [1:] 4. Aufl. [2–3:] 3. Aufl., Berlin, 1959. Translated by Gilbert Highet, *Paideia: The Ideals of Greek Culture* (NY: Oxford University Press, 1939–1944).

5. See Pseudo Dionysius, *The Divine Names* in *The Complete Works* (New York: Paulist, 1987), 25–34.

6. Franz Rosenzweig. *The Star of Redemption* (Wisconsin: University of Wisconsin Press, 2005), 319.

7. Ibidem.

8. Novalis. *Das allgemeine Brouillon, Materialien zur Enzyklopädistik 1798/99, Nr. 85,* translated by David W. Wood. *Notes for a Romantic Encyclopedia. Das allgemeine Brouillon* (New York: SUNY Press, 2007).

9. See Thomas Wolfe, *You Can't Go Home Again*, with an introduction by Edward C. Aswell (New York: Scribner, 1968).

10. Werner Hamacher, "En-counterings of Time," in *History Today*, special issue of *Philosophy Today* 60, no. 4 (Fall 2016), ed. Jean-Luc Nancy and Marcia Sá Cavalcante Schuback (Chicago: De Paul University), 839–51.

11. See Marcus Terentius Varro, *On the Latin Language.* [2], Books VIII–X Fragments (Cambridge MA: Harvard University Press, 1951).

12. "First, forms exist which really are in a sense halfway between noun and verb, belonging half with the one and half with the other. This is true of participles, infinitives, and gerunds." See Jakob Wackernagel, *Lectures on Syntax: With Special Reference to Greek, Latin, and Germanic* (Oxford: Oxford University Press, 2009), 98.

13. L. Horton-Smith, "The Origin of the Gerund and Gerundive," in *The American Journal of Philology* 15, no. 2 (1894), 194–216.

14. About this difficulty, see Jakob Wackernagel. *Lectures on Syntax: With Special Reference to Greek, Latin, and Germanic* (Oxford: Oxfrod University Press, 2009).

15. For a "linguistic vagabondage" on the road of the gerundive in different languages and peoples, see Jean-Pierre Minaudier. *Poésie du Gérondif (vagabondages linguistiques d'un passion de peuples et de mots)* Le Tripode, 2014.

16. Samuel Weber, *Theatricality as Medium* (New York: Fordham University Press, 2004), 15–16. I discovered Weber's discussion about the gerund and the present participle when I was proofreading this manuscript to send for publication. That is the main reason for not having engaged more with Weber's book above all in relation to Heidegger.

17. Pascal Quignard, "Note sur le gérondif latin," in *Sur l'image qui manque à nos jours* (Paris: Arléa, 2014), 45–48.

18. Samuel Weber, *Theatricality as Medium*, op. cit., 16.

19. Pascal Quignard, op. cit., 46.

20. Aristotle, *Metaphysics*. [1], Books I–IX (Cambridge, MA: Harvard University Press, 1933), "For as the eyes of bats are to the blaze of day, so is the reason in our soul to the things which are by nature most evident of all."

21. Here we could even speak about exile as an ontological transit camp, when taking this expression in the way the German film artist Harun Farocki has done. See Harun Farocki. *Aufschub. Respite.* https://www.youtube.com/watch?v=2ciiA0bySPc.

22. Hannah Arendt, "What Remains? The Language Remains: A Conversation with Günther Gaus," in *Essays in Understanding, 1930–1954: Formation, Exile and Totalitarianism* (New York: Chock Books, 2005); see also Jacques Derrida's comments on this thought by Arendt in *Le monologinguisme de l'autre ou La Protèse d'origine* (Paris: Galilée, 1996) translated by Patrick Mensah as *Monolinguism of the Other or the Prosthesis of Origin* (Stanford: Stanford University Press, 1998).

23. Marc Robinson (ed.), *Altogether Elsewhere: Writers on Exile* (Winchester, MA: Faber and Faber, 1994).

24. See Werner Hamacher, "Affirmative, Strike," 13 *Cardoso Law Review* (1991) 9, 1133. "*Affirmative* is not *aformative*, afformance 'is' the event of forming, itself formless, to which all forms and all performative acts remain exposed. (The Latin prefix *ad-* and accordingly *af-*, marks the opening of an act, and of an act of opening, as in the very appropriate example of *affor*,

meaning 'addressing,'" e. g. when taking leave.) But of course, in affirmative one must also read *aformative*, as determined by affirmative."

25. Paul Valéry, "Poésie et pensée abstraite," in *Oeuvres* I (Paris: Gallimard, 1957), 1314–39.

26. In Peter Szondi's readings of Paul Celan's poem *Engführung*, one can find a similar attempt to read the poem as the writing of its own reading. See Peter Szondi, "Reading 'Engführung': An Essay on the Poetry of Paul Celan," in *Boundary 2*, 11, no. 3, *The Criticism of Peter Szondi* (Spring, 1983), pp. 231–64 (Duke University Press).

Chapter 1

1. Rancière, Jacques, *En quel temps vivons-nous?* (Paris: La Fabrique èditions, 2017).

2. Jean-Luc Nancy, *Que faire?* (Paris: Galilée, 2016).

3. For a discussion about being with the without, see Schuback, Marcia Sá Cavalcante & Nancy, Jean-Luc (eds.) *Being with the Without* (Stockholm: Axl Books, 2013), and Marcia Sá Cavalcante Schuback, "Being without (Heidegger)" in *Gatherings*. Papers from the 50th Anniversary Conference. The Heidegger Circle Annual, volume 7, 2017, 70–84.

4. Husserl E. Husserliana: Gesammelte Werke. Bd 29, *Die Krisis der europäischen Wissenschaften und die transzendentale Phänomenologie, Ergänzungsband, Texte aus dem Nachlass 1934–1937* (Dordrecht: Kluwer; 1993), translated by David Carr, *The Crisis of European Sciences and Transcendental Phenomenology: An Introduction to Phenomenological Philosophy* (Evanston, IL: Northwestern University Press, 1978 [1970]).

5. Herbert Marcuse, *One-dimensional Man: Studies in the Ideology of Advanced Industrial Society* (Boston: Beacon, 1991 [1964]).

6. Aristotle, *Metaphysics*, op. cit., 1072a.

7. This formulation is a paraphrase of Schelling's phrase in the *Erlangen Lectures* that reads: "To go through all things and to be nothing, namely, to be nothing such that it could always be otherwise—this is the demand," in Friedrich Schelling, *Initia Philosophiae Universea: Erlanger Vorlesung WS 1820/21* (Bonn: Bouvier, 1969), 10.

8. Philippe Lacoue-Labarthe Typographie (Paris: Aubier-Flammarion, 1975), edited by Christopher Fynsk, *Typography* (Stanford: Stanford University Press, 1989).

9. Aristotle, *The Physics*. [1], Books I–IV (Cambridge, MA: Harvard University Press, 1957) and [2], Books V–VIII (Cambridge, MA: Harvard University Press, 1934).222b.

10. On this understanding of status quo, see Fredric Jameson, *The Seeds of Time* (NY: Columbia University Press, 1994).

11. About the concept of stasis as civil war in ancient Greece, see Nicole Loraux, *La cité divisée* (Paris: Payot, 1997b), translated by Corine Pache and Jeff Fort, *The Divided City: On Memory and Forgetting in Ancient Athens* (New York. Zone Books, 2002). For a political-philosophical discussion of the concept, see Hannah Arendt, *On Revolution* (New York: Viking, 1963), and Giorgio Agamben. *La guerre civile. Pour une théorie politique de la stasis* (Paris: Éditions Points, 2015).

12. Jean-François Lyotard, *La condition postmoderne: rapport sur le savoir* (Paris: Éd. de minuit, 1979), translated by Geoff Bennington and Brian Massumi, *The Postmodern Condition: A Report on Knowledge* (Manchester: Manchester University Press, 1984).

13. Michel Foucault, *Les mots et les choses: une archéologie des sciences humaines* (Paris: Gallimard,1995), eng. *The Order of Things: An Archaeology of the Human Sciences* (London: Repr., Tavistock/Routledge, 1989 [1974]).

14. Bill Ashcroft, Gareth Griffiths, and Helen Tiffin, *Postcolonial Studies: The Key Concepts* (London: Routledge, 2013); Robert J. C. Young, *Postcolonialism: An Historical Introduction* (Oxford. Blackwell, 2001); Gayatri Chakravorty Spivak, *A Critique of Postcolonial Reason: Toward A History of the Vanishing Present* (Cambridge, MA: Harvard University Press, 1999); Bhabha, Homi K., *The Location of Culture* (London: Routledge, 1994).

15. Boris Grojs, Anne von der Heiden, and Peter Weibel (eds.), *Zurück aus der Zukunft: osteuropäische Kulturen im Zeitalter des Postkommunismus* (Frankfurt am Main: Suhrkamp, 2005).

16. Lyotard, op. cit.

17. Gerhard Richter, *Afterness: Figures of Following in Modern Thought and Aesthetics* (New York: Columbia University Press, 2011).

18. Jean-François Lyotard, "Foreword: After the Words," in Joseph Kosuth, *Art after Philosophy and After: Collected Writings, 1966–1990* (Cambridge, MA: MIT Press, 1991), xv.

19. On Gadamer's concept of *Verweilen*, see H-G Gadamer, "Wort und Bild – 'so wahr, so seiend'" (1992) in *Gadamer Lesebuch* (Tübingen: Mohr Siebeck, 1997), 172–200, English translation, *Artworks in Word and Image: So True, So Full of Being!* (Goethe) (1992), in *Theory, Culture and Society*, January 2006 (Goldsmiths, University of London, UK).

20. Near-being translates literally the Swedish word for presencing, *närvarande, när*, meaning near, and *varande*, being.

21. Aristotle, *Peri mnemes kai anamneseos, De memoria et reminiscentia*, in *On the Soul. Parva Naturalia. On Breath.* Translated by W. S. Hett. Loeb Classical Library 288 (Cambridge, MA: Harvard University Press, 1957), 449b15.

22. For a discussion about this distinction, see Paul Ricoeur, *La mémoire, l'histoire, l'oubli* (Paris: Seuil, 2000), translated by Kathleen Blamey and David

Pellauer, *Memory, History, Forgetting* (Chicago: University of Chicago Press, 2004), 43–58.

23. Jacques Derrida, "Dialanguages," in *Points . . . Interviews, 1974–1994*, translated by Peggy Kamuf et al. (Stanford: Stanford University Press, 1995), 132–55.

24. Ibidem, p. 143–44.

25. Maurice Halbwachs, *La mémoire collective* (Paris: PUF, 1950), 15, translated by Lewis A. Coser, *On Collective Memory* (Chicago: Chicago University Press, 1992).

26. See Edmund Husserl, *Phantasie, Bildbewusstsein, Erinnerung. Zur Phänomenologie der anschaulichen Vergegenwärtigungen. Texte aus dem Nachlass (1898–1925). Husserliana* bd 23 (The Hague: Martinus Nijhoff, 1980); English translation *Phantasy, Image Consciousness, and Memory (1898–1925), Collected Works*, vol. 11 (Netherlands: Springer, 2005).

27. Aristotle, *Peri mnemes kai anamneseos, De memoria et reminiscentia*, in *On the Soul, Parva Naturalia*, op. cit., 449b15.

28. Aristotle, ibidem.

29. See here the work of John Sallis, *Force of Imagination: The Sense of the Elemental* (Bloomington: Indiana University Press, 2000).

30. Diderot, *Élements de phyisologie* (Paris: Didier, 1964).

31. Novalis, *Das allgemeine Brouillon*, op. cit., 355.

32. Thomas Hobbes, *Leviathan* (New York: Norton, 1997).

33. Derrida, Jacques, "Dialanguages," op. cit., 143.

34. St. Augustine, *Confessions*, translated by Henry Chadwick (London: Oxford University Press, 1991), book 10, 193.

35. See Paul Ricoeur, *Memory, History, Forgetting*, op. cit.

36. Vladimir Nabokov, *Speak, Memory. An Autobiography Revisited* (London: Penguin Books, 1967).

37. Edward Said *Reflections on Exile and Other Literary and Cultural Essays* (London: Granta, 2000), 186.

38. Nabokov, op. cit., 29.

39. Ibidem.

40. Ibidem, 22.

41. Heraclitus, fragment 122.

Chapter 2

1. See Heidegger's discussions on *Heimatlosigkeit* a. O in "Zur Seinsfrage," *Wegmarken, Gesamtausgabe* 9 (Frankfurtam Main: Vittorio Klostermann, 1978), 381, edited by William Mc Neill, *Pathmarks* (Cambridge: Cambridge University Press; 1998), in "Bauen, Wohnen, Denken," *Vorträge und Aufsätze*,

Gesamtausgabe, henceforth GA 7 (Frankfurt am Main: Vittorio Klostermann, 2000), 163, translated by David Farell Krell, *Basic Writings* (London: Taylor & Francis, 2010), and "Brief über den Humanismus" in *Wegmarken*, GA 9, op. cit., 338. See also Felicetti Ricci, "Gagner la Heimatlosigkeit," in *Heidegger Studies*, vol. 24 (Berlin: Duncker & Humblot, 2008), 61–102.

2. Martin Heidegger, *Sein und Zeit*, GA 2 (Frankfurt am Main: Vittorio Klostermann, 1977), 170, henceforth SZ, translated by Joan Stambaugh as *Being and Time* (NY: State University of New York Press, 1996), henceforth BT.

3. Ibidem.

4. Ibidem, SZ, 170, BT, 159.

5. Martin Heidegger, Die *Grundbegriffe der Metaphysik. Welt – Endlichkeit – Einsamkeit*, GA 29–30 (Frankfurt am Main: Vittorio Klostermann, 1983), translated by William McNeill and Nicholas Walker as *The Fundamental Concepts of Metaphysics: World, Finitude, Solitude* (Bloomington and Indianapolis: Indiana University Press, 1995).

6. Heidegger, SZ, 229, BT, 210.

7. Ibidem, SZ, 88.

8. "The inner relationship of my own work to the Black Forest and its people comes form a centuries-long and irreplaceable rootedness in the Alemannian-Swabian soil," in "Why do I stay in the Provinces?" (1934), English version published in Thomas Sheehan, *Heidegger: The Man and the Thinker* (Chicago: Precedent, 1981), 28.

9. See Martin Heidegger, *The Question concerning Technology and Other Essays*, translated by William Lovitt (New York: Harper Torchbooks 1977). Samuel Weber also proposed "emplacement"; see *Theatricality as Medium*, op. cit., 59.

10. See Jean-Luc Nancy, "L'existence exile," in *Cahiers Intersignes*, ed. Fethi Benslama, 14–15 (2001); Alejandro Vallega, *Heidegger and the Issue of the Space: Thinking on Exilic Grounds* (Pennsylvania: Pennsylvania State University Press, 2003); and Donatella di Cesare's article "Exile: Human Condition in the Globalized World," in *Philosophy Today*, SPEP Supplement, 2008, 85–93. For a discussion of Heidegger and Nancy on the topic of existence in exile, see my "Exile and Existential Disorientation," in *Dis-orientations. Philosophy, Literature and the Lost Grounds of Modernity*, co-ed with Tora Lane (London: Rowman & Littlefield, 2015).

11. "Zeitlichkeit ist das ursprüngliche, 'Außer-sich an und für sich selbst," Heidegger, SZ, 329, BT, 302.

12. See my article "Die Gabe und Aufgabe des Währenden," in *Translating Heidegger's Sein und Zeit*, edited by Christian Ciocan, *Studia Phenomenologica* 5, no. 205, 201–14.

13. "Das Sein selbst, zu dem das Dasein sich so oder so verhalten kann und immer irgendwie verhält, nenne wir Existenz. Und weil die Wesensbestimmung dieses Seienden nicht durch Angabe eines sachhaltigen Was

vollzogen werden kann, sein Wesen vielmehr darin liegt, daß es je sein Sein als seiniges zu sein hat, ist der Titel Dasein als reiner Seinsausdruck zur Bezeichnung dieses Seienden gewählt," *SZ*, 12. English version: "We shall call the very being to which Da-sein can relate in one way or another, and somehow always does relate, existence [*Existenz*]. And because the essential definition of this being cannot be accomplished by ascribing to it a "what" that specifies its material content, because its essence lies rather in the fact that it in each instance has to be its being as its own, the term Dasein, as a pure expression of being, has been chosen to designate this being." *BT*, 10.

14. *Die Frage der Existenz ist immer nur durch das Existieren selbst ins Reine zu bringen.* SZ, 12, BT, 10.

15. *Daher kann sich das Dasein existierend nie als vorhandene Tatsache feststellen, die »mit der Zeit« entsteht und vergeht und stückweise schon vergangen ist. Es »findet sich« immer nur als geworfenes Faktum.* SZ, 328, BT, 301.

16. "Die existenziale Analytik ihrerseits aber ist letztlich existenziell, d. h. ontisch verwurzelt. Nur wenn das philosophischforschende Fragen selbst als Seinsmöglichkeit des je existierenden Daseins existenziell ergriffen ist, besteht die Möglichkeit einer Erschließung der Existenzialität der Existenz und damit die Möglichkeit der Inangriffnahme einer zureichend fundierten ontologischen Problematik überhaupt. Damit ist aber auch der ontische Vorrang der Seinsfrage deutlich geworden." SZ, 13, BT, 18.

17. Jacques Derrida, *Heidegger: La question de l'être et de l'histoire. Cours de l'ENS-Ulm (1964–65)* (Paris: Galilée, 2013), translated by Geoffrey Bennington, *Heidegger: The Question of Being and History* (Chicago: University of Chicago Press, 2016).

18. Emily Dickinson, poem 812, "A Light Exists in Spring," in *The Complete Poems of Emily Dickinson*, edited by Thomas H. Johnson (Boston/Toronto: Little, Brown and Company, 1960), 395.

19. SZ, 314, BT, 290.

20.

Existierend kommt es nie hinter seine Geworfenheit zurück, so daß es dieses »daß es ist und zu sein hat« je eigens erst aus *seinem* Selbstsein entlassen und in das Da führen könnte. Die Geworfenheit aber liegt nicht hinter ihm als ein tatsächlich vorgefallenes und vom Dasein wieder losgefallenes Ereignis, das mit ihm geschah, son-dern das *Dasein ist* ständig – solange es ist – als Sorge sein »Daß«. *Als dieses Seiende*, dem überantwortet es einzig als das Seiende, das es ist, existieren kann, *ist es existierend* der Grund seines Seinkönnens. Ob es den Grund gleich *selbst nicht* gelegt hat, ruht es in seiner Schwere, die ihm die Stimmung als Last offenbar macht. SZ, 284, BT, 262

21. "Es hat nicht ein Ende, an dem es nur aufhört, sondern existiert endlich." SZ, 329, BT, 303.

22. Cf. E. Benvéniste, Le système sublogique des prépositions en latin, in *Recherches structurales*, Publ. à l'occasion du cinquantenaire de M. Louis Hjelmslev (Copenhague: Cercle linguistique de Copenhague. Travaux. Vol. 5, 1949), and H. Maldiney, *Aîtres de la langue et demeures de la pensée* (Lausanne: L'Âge d'Homme, 1975).

23. SZ, 132, BT, 124.

24. SZ, 350, BT, 321.

25. "ursprüngliche "Außer-sich" and und für sich selbst." SZ, 329, BT, 302.

26. *Wenn ihm (Dasein) sonach in irgendeiner Weise Räumlichkeit zukommt, dann ist das nur möglich auf dem Grunde dieses In-Seins.* SZ, 104–5, BT, 97.

27. *Dessen Räumlichkeit aber zeigt die Charactere der Ent-fernung und Ausrichtung.* SZ, 105, BT, 97.

28. Following here is the vocabulary used by Oscar Becker in his phenomenology of "oriented space" Oscar Becker, "Beiträge zur phänomenologische Begründung der Geometrie und ihrer physikalischen Anwendungen" in *Jahrbuch für Philosophie und phänomenologuischen Forschung*, Bd VI, 1923. See Heidegger's discussions related to Becker's investigations and to the topic in SZ, sections 23, 24.

29. Immanuel Kant, "Was heißt: Sich im Denken orientieren?" (1786), in *Werke*. Akademie Ausgabe VIII, s. 131–47, translated by H. B. Nisbet, "What Is Orientation in Thinking? in Immanuel Kant," *Political Writings* (Cambridge: Cambridge University Press, 1991). See also Heidegger's interpretation of this text by Kant in SZ, section 23.

30. *Im Dasein liegt eine wesentliche Tendenz auf Nähe.* SZ, 105, BT 98.

31. *Das Dasein ist zwar ontisch nicht nur nahe oder gar das nächste – wir sind es sogar je selbst. Trotzdem oder gerade deshalb ist es ontologisch das Fernste.* SZ, 21, BT, 18.

32. SZ, 146, BT, 137.

33. Ibidem.

34. "Er (der Ausdruck 'Sicht') entspricht der Gelichtetheit, als welche wir die Erschlossenheit des Da charakterisierten." SZ 147, BT 137.

35. Ibidem.

36. Jacques Derrida, *Le toucher, Jean Luc Nancy* (Paris: Galilée, 2000), translated by Christine Irizarry as *On Touching* (Stanford: Stanford University Press, 2005).

37. The phenomenon of care in its totality is essentially something that cannot be split up; thus, any attempts to derive it from special acts or drives such as willing and wishing or urge and predilection, or of constructing it out of them, will be unsuccessful. SZ, 192, BT, 181.

38. SZ, 192, BT, 180.

39. SZ, 390, BT, 357.

40. "Als Sorge ist das Dasein das Zwischen," SZ, 374, BT, 343; note that the ist is emphasized.

41. SZ, 325, BT, 299.

42. Emmanuel Levinas, "L'existentialisme, l'angoisse et la mort," in Exercices de la patience. Heidegger, nos. 3–4 (Paris: Obsidiane, 1982).

43. The way Heidegger understands the "to-come" basically in relation to the claim to overcome [metaphysics] should be distinguished from Derrida's thoughts on the "to-come." This difference appears on the basis of their diametric views on the "end," "the end of the world," between Heidegger's apocalyptism and Derrida's messianim.

44. SZ, 39, BT, 34.

45. Das Geschehen als Wirkung anzusetzen: und die Wirkung als Sein; das ist der doppelte Irrthum, oder Interpretation, deren wir uns schuldig machen, Also z. B. "der Blizt leuchtet" - : "leuchten" ist ein Zustand an uns; aber wir nehmen ihn nicht als Wirkung auf uns, und sagen: "etwas Leuchtendes" als ein "An-sich und suchen dazu ein Urheber, den Blitz." Friedrich Nietzsche, Nachgelassene Fragmente 1885–1887, Kritische Studienausgabe, 2232, 104.

46. Heidegger uses this term twice in Being and Time. SZ, section 10, 47, and section 83, 473.

47. See Peter Trawny, Heidegger & the Myth of a Jewish World Conspiracy (Chicago: University of Chicago Press, 2015); Peter Trawny, Freedom to Fail: Heidegger's Anarchy (John Wiley & Sons, 2015); and David Krell, Ecstasy, Catastrophe, op. cit.

48. "Unbedingten Anthropomorphie." Martin Heidegger, Nietzsche II, GA 6.2 (Frankfurt am Main: Vittorio Klostermann, 1997), 20, translated by David Farell Krell, Nietzsche (San Francisco: Harper & Row, 1979).

49. Martin Heidegger, "Der Spruch des Anaximanders" in Holzwege, GA 5 (Frankfurt am Main: Vittorio Klostermann, 1977), 327, translated by Julian Young and Kenneth Haynes, Off the Beaten Track (Cambridge: Cambridge University Press, 2002).

50. It is in the realm of these questions that Heidegger's political engagement and his "onto-historical" or "metaphysical" antisemitism emerges. See P. Trawny, and A. Mitchell, Heidegger, die Juden, nocheinmal (Frankfurt am Main: Vittorio Klostermann, 2015).

51. Martin Heidegger, Beiträge zur Philosophie, GA 65 (Frankfurt am Main: Vittorio Klostermann, 3 ed. 2003) translated by Richard Rojcewicz and Daniela Vallega-Neu, Contributions to Philosophy, hereafter CP (Indiana: Indiana University Press, 2012), §239, 294.

52. Martin Heidegger, Nietzsche II, GA 6.2, op. cit., 20.

53. "Der Zeit-Raum ist die ereignete Erklüftung der Kehrungsbahnen des Ereignisses, der Kehre zwischen Zugehörigkeit und Zuruf, zwischen Seinsverlassenheit und Erwinkung (das Erzittern der Schwingung des Seyns selbst!). Nähe und Ferne, Leere und Schenkung, Schwung und Zögerung, all dieses darf

nicht zeitlich-räumlich begriffen werden von den üblichen Zeit und Raum-Vorstellungen her, sondern umgekehrt, in ihnen liegt das verhüllte Wesen des Zeit-Raumes." GA 65, sections 239, 372. English version: "Time-space is the appropriated sundering of the turning paths of the event, the sundering of the turning between belonging and call, between abandonment by being and beckoning intimation (the trembling in the oscillation of beyng itself!). Nearness and remoteness, emptiness and bestowal, verve and hesitation—in these the hidden essence of time-space resides, so they cannot be grasped temporally and spatially on the basis of the usual representations of time and space." CP, section 239, 294.

54. Martin Heidegger, GA 65, CP, section 242, 301.

55. Martin Heidegger, "Die Kehre," in Bremer und Freiburger Vorträge, GA 79 (Frankfurt am Main: Vittorio Klostermann, 1994), translated by Andrew Mitchell as Bremen and Freiburg Lectures: Insight into That Which Is and Basic Principles of Thinking (Bloomington: Indiana University Press, 2012).

56. Ibidem.

57. Martin Heidegger "Brief über den Humanismus," in Wegmarken, GA 09, 342; "Zeit und Sein," in Zur Sache des Denkens, GA 14 (Frankfurt am Main: Vittorio Klostermann, 1962–64), 14.

58. A suggestion made by Samuel Weber, op. cit.

59. Françoise Dastur, Dire le temps (La Versanne: Encre marine, 2002), 81.

60. See Heidegger's own criticism above all in Beiträge zur Philosophie, GA 65, 301, and the forthcoming volume of the Gesamtausgabe that is being edited by F. v Hermann under the title given by Heidegger "Laufende Anmerkungen zu »Sein und Zeit«."

61. Martin Heidegger, GA 65, CP, sections 238–42.

62. This question entitles the book by Nikolai Chernyshevsky, re-asked by Lenin and more recently rephrased by Jean-Luc Nancy in terms of "Que faire? Op. cit.

63. Martin Heidegger, Was heißt Denken? GA 08 (Frankfurt am Main: Vittorio Klosterman (1951–52/2002), translated by J. Glenn Gray, What Is Called Thinking? (New York: HarperCollins, 1976).

64. Martin Heidegger, "Die Frage nach der Technik," in Vorträge und Aufsätze, GA 07 (Frankfurt am Main: Vittorio Klostermann, 2000), translated by Willem Lovitt, The Question concerning Technology and Other Essays, op. cit.

65. Ibidem.

66. Ibidem.

67. Martin Heidegger, "Zeit und Sein" in Zur Sache des Denkens, GA 14, op. cit., translated by Joan Stambaugh as "Time and Being" (New York, Harper & Row, 1972), 22.

68. Ibidem.

69. Martin Heidegger, *Identität und Differenz* (Pfullingen: Günther Neske, 1957).

70. Martin Heidegger, Gesellscahft, *Jahresgabe* 2011/2012.

71. Martin Heidegger, *Zu eigenen Veröffentlichungen*, GA 82 (Frankfurt am Main: Vittorio Klostermann, 2018).

72. For a discussion about Heidegger's attention to the present participle, see Samuel Weber, op. cit., 17–22.

73. Martin Heidegger, *Metaphysische Anfangsgründe der Logik im Ausgang von Leibniz*, GA 26 (Frankfurt am Main: Vittorio Klostermann, 1978), 242–43, translated by Michael Heim as *The Metaphysical Foundations of Logic* (Bloomington: Indiana University Press, 1992), 188.

74. Ibidem, 137.

75. Ibidem.

76. Ibidem.

77. Ibidem.

78. Ibidem.

79. "Während wir soeben dem Sein andachten, hat sich erwiesen: das Eigentümliche des Seins, das, wohin es gehört und worin es einbehalten bleibt, zeigt sich im Es gibt und dessen Geben als Schicken." *Zeit und Sein*, op. cit., 10, *Time and Being*, op. cit., 10.

80. Aristotle, *Peri mnemes kai anamneseos, De memoria et reminiscentia*, in On the Soul. *Parva Naturalia*, op. cit., 449b15.

81. For an account on Heidegger's understanding of nearness, see Krzysztof Ziarek, *Inflected Language: Towards a Hermeneutics of Nearness: Heidegger, Levinas, Stevens, Celan* (Albany: State University of New York Press, 1994), and Gerhard Richter, *Afterness: Figures of Following in Modern Thought and Aesthetics* (New York: Columbia University Press, 2011).

82. In the insightful reading of the late Heidegger, and specially of the thought of the fourfold, Andrew J. Mitchell draws attention to the figure of the "while," mainly in Heidegger's readings of Hölderlin. See Andrew J. Mitchell, *The Fourfold: Reading the Late Heidegger* (Evaston, IL: Northwestern University Press, 2015), 280–98.

83. Martn Heidegger, *Was ist das – die Philosophie?* (Pfullingen: Günther Nske, [1956], 1976), 13.

Chapter 3

1. Beckett uses the expression "issueless," meaning "without exit," in a short piece written first in French under the title *Sans*, which he translated into English as *Lessness*. See Samuel Beckett, *Sans* [*Lessness*] in Van Hulle, Dirk. "Sans." *The Literary Encyclopedia*, first published 01 March 1, 2004.

[http://www.litencyc.com/php/sworks.php?rec=true&UID=2307, accessed 26 July 2017.]

2. Andrew Mitchell writes the following about his translation of *Verwindung* as conversion:

> The term is presented as an alternative to the history of meta-
> physics and as a relationship to pain. We know from elsewhere
> that it is precisely not a matter of "overcoming" (*Überwindung*)
> that is at stake, but instead something else. In the *Verwindung*,
> the prevailing situation (that which is) is seen in a flash to be
> a dispensation of beyng, dislodging it of any presumed stability.
> *Verwindung* can thus be heard as "bringing to a turning point"
> or pivot point. It is the moment that the limit is achieved and
> what once was construed as 'inside' shows itself now as exposed
> to an outside lying beyond it. One achieves a position at the
> limit (of metaphysics, of beings, of being itself) around which
> the whole will revolve. A new constellation becomes visible now
> in a change of philosophical seasons. The perspective from the
> limit that is able to see how metaphysics is a dispensation (i.e.,
> is sent, has an outside) is now said to have "converted" that
> former position. But just as a pain is converted into a part of
> one's identity through the formation of a scar, so too are there
> traces of metaphysics to be found here as well. There is no
> complete *Verwindung*, no final conversion, for such would only
> be another 'overcoming' (*Überwindung*), as these are endemic to
> metaphysics. *Verwindung* in German carries none of the religious
> overtones of "conversion" in English (German, *Umkehrung*), but
> as the discussion is of lightning flash moments of the arrival of
> grace, these overtones are not entirely foreign to the tenor of
> the text." Andrew Mitchell, "Translator's Foreword," in *Martin
> Heidegger: Bremen and Freiburg Lectures: Insight into That Which
> Is and Basic Principles of Thinking* (Bloomington and Indianapolis:
> Indiana University Press, 2012), xii

Despite this explanation, I will use "enduring" to translate *verwinden* and *Verwindung* for the sake of stressing the holding on [the suffering] until it goes over by itself.

3. Martin Heidegger, *Was heißt Denken?* op. cit., *What Is Called Thinking?*, op. cit.

4. Ibidem, 17.

5. Ibidem.

6. Ibidem.

Once we are so related and drawn to what withdraws, we are drawing into what withdraws, into the enigmatic and therefore mutable nearness of its appeal. Whenever man is properly drawing that way, he is thinking—even though he may still be far away from what withdraws, even though the withdrawal may remain as veiled as ever. All through his life and right into his death, Socrates did nothing else than place himself into this draft, this current, and maintain himself in it. This is why he is the purest thinker of the West. This why he wrote nothing. For anyone who begins to write out of thoughtfulness must inevitably be like those people who run to seek refuge from any draft too strong for them. An as yet hidden history still keeps the secret why all great Western thinkers after Socrates, with all their greatness, had to be such fugitives. Thinking has entered into literature; and literature has decided the fate of Western science which, by way of the *doctrina* of the Middle Ages, became the *scientia* of modern times. In this form all the sciences have leapt from the womb of philosophy, in a twofold manner. The sciences come out of philosophy, because they have to part with her.

7. Jacques Derrida, *Edmund Husserl, L'origine de la géométrie* (Paris: PUF, 1962), translated by John P. Leavey Jr., *Edmund Husserl's Origin of Geometry* (Lincoln: University of Nebraska Press, 1989 [1978]).

8. The attention that Blanchot dedicated to the question of the between also endorses this view. In his discussions of Bataille in *La Communauté Inavouable: The Unavowable Community*, for instance, Blanchot reads Bataille as a thinker and writer of the between continuity and discontinuity, distancing thereby from current views who see in Bataille rather a thinker of continuity. The between is comprehended as "passage," as "stepping," a sense to which Blanchot will always insist. See, for example, *La Communauté inavouable* (Paris: Les Éditions de Minuit, 1983), 42, translated by Pierre Joris, *The Unavowable Community* (Barrytown, NY: Station Hill Press, 1988).

9. "*Fiction théorique*" is an expression used by Blanchot when discussing the inventions of grammarians such as Franz Bopp, the father of comparative grammar. See Maurice Blanchot, *L'Écriture du Désastre* (Paris: Gallimard, 1980), 166, translated by Ann Smock (Lincoln: University of Nebraska Press, 1995).

10. See Jacques Derrida, "Demeure," in Maurice Blanchot, *L'instant de ma mort* (Paris: Gallimard, 2002 [1994], translated by Elizabeth Rottenberg as *The Instant of My Death* / Maurice Blanchot, published together with Jacques Derrida, Demeure: Fiction and Testimony (Stanford: Stanford University Press, 2000), 89.

11. For a discussion on Blanchot and the question of being and of the neuter, see Marlène Zarader, *L'être et le neutre. À partir de Maurice Blanchot* (Lagrasse, editions Verdier, 2001).

12. For a very good analysis of Blanchot's thoughts on endless end, see Aïcha Liviane Messina, "The Apocalypse of Blanchot," in *Philosophy Today*, special issue *History Today* (Chicago: De Paul University, 2016), 877–93.

13. Leslie Hill. *Maurice Blanchot and Fragmentary Writing: A Change of Epoch* (London/NY: Continuum, 2012), 175.

14. Ibidem.

15. Christopher Fynsk, *Last Steps: Maurice Blanchot's Exilic Writing* (New York: Fordham University Press, 2013).

16. Ibidem, 4.

17. Leslie Hill, op. cit., 30.

18. Christopher Fynsk, op. cit., 126.

19. Maurice Blanchot, *L'Entretien Infini* (Paris: Gallimard, 1969), *The Infinite Conversation* (Minneapolis and London: University of Minnesota Press, 2013), 307.

20. Ibidem, 308.

21. Ibidem.

22. Ibidem, 152.

23. Ibidem.

24. Ibidem, 158.

25. Ibidem, 167.

26. I want to thank Laura Marin for sending me her inspiring article on "Survivances du neutre," in which she presents her discovery on the medical concept of the neuter during the Renaissance and how its medical signification is actualized in contemporary debates on the neuter, particularly in Blanchot and Barthes. According to Marin, the medicine in the Renaissance assumed the neuter as a category of life, a life that is neither sick nor healthy, expressing the passage and transition from one state to the other through a succession of degrees. In the medical debates on the neuter, it the need to leave behind the dualistic logic of either and understand life more as a process and progression. See Laura Marin, "Survivances du neuter," in *Pour une politique du résiduel en literature, Cahiers Echinox*, vol. 33 (Cluj-Napoca, Romània, 2017).

27. Maurice Blanchot, *Le Pas au-delà* (Paris: Gallimard, 1973), 97, translated by Lycette Nelson, *The Step Not Beyond* (New York: State University of New York Press), 69.

28. Jean-Marie Pontévia, *Tout a peut-être commence par la beauté. Écrits sur l'art et pensées detachées* (Paris: William Blake & Co. Édit. 1985), 218.

29. Maurice Blanchot, *L'Écriture du désastre* (Paris: Gallimard, 1980), 58, translated into English by Ann Smock, *The Writing of the Disaster* (Lincoln and London: University of Nebraska Press, 1995), 33.

30. Leslie Hill, op. cit., 31.

31. Maurice Blanchot, *L'Écriture du désastre*, op. cit., 58, English translation, *The Writing of the Disaster*, op. cit., 34.

32. Ibidem, 36, English, 18.

33. Ibidem, 99, English, 60.

34. Ibidem, 100, English, 60–61.

35. Ibidem, 11, English, 3.

36. Ibidem, 7, English, 1.

37. Maurice Blanchot. *Thomas l'obscur* (Paris: Gallimard, 1950), translated by Lydia Davis, Paul Auster, and Robert Lamberton as "Thomas the Obscure," in *The Station Hill Blanchot Reader*, ed. Goerge Quasha (Barrytown, NY: Station Hill, 1998).

38. Maurice Blanchot, *Faux Pas* (Paris: Gallimard, 1943), translated by Charlotte Mandelas as *Faux Pas* (Stanford: Stanford University Press, 2001).

39. Maurice Blanchot, *Pau au-delà*, op. cit., *The Step Not Beyond*, op. cit.

40. Maurice Blanchot, *L'instant de ma mort*, op. cit., *The Instant of My Death*, op. cit.

41. Maurice Blanchot, *Le Pas au-delà*, op. cit., 8, *The Step Not Beyond*, op. cit., 1.

42. Ibidem.

43. Ibidem, 145, English, 105.

44. Martin Hägglund, *Radical Atheism: Derrida and the Time of Life* (Stanford: Stanford University Press, 2008), 3.

45. Christopher Langlois, "Temporal Exile in the Time of Fiction: Reading Derrida Reading Blanchot's *The Instant of my Death*," in *Mosaic* 48, no. 4 (2015): 17–32. Special Issue: *A Matter of Life/Death*.

46. Emmanuel Levinas, op. cit., 85.

47. Emmanuel Levinas, "Reality and Its shadow," in *Unforseen History* (Urbana: University of Illinois Press, 2004).

48. Maurice Blanchot, "Literature and the Right to Death," in *The Work of Fire*, translated by Charlotte Mandel (Stanford: Stanford University Press, 1995).

49. Michel Foucault, *The Thought from Outside*, and Maurice Blanchot, *Michel Foucault as I Imagine Him* (NY: Zone Books, 1990), 15.

50. Ibidem.

51. Maurice Blanchot, *L'écriture du désastre*, op. cit., 100.

52. See Emmanuel Levinas, *Autrement qu'être ou au-delà de l'essence* (Netherlands: Springer 1974), translated by Alphonso Lingis, *Otherwise than Being, or Beyond Essence* (Pittsburgh: Duquesne University Press, 1998–[1974]), 180. For a discussion of the inside-out in Levinas, see Ramona Rat, *Un-common Sociality: Thinking Sociality with Levinas* (Stockhlm: Södertörn Philosophical Studies 19, 2016).

53. Maurice Blanchot, *Le Pas au-delà*, op. cit., 49, *The Step Not Beyond*, op. cit., 33.

54. Ibidem, 9, English, 2.

55. For a discussion about this dialectic of uprootedness and taking roots anew as ground of modernity, see Marcia Sá Cavalcante Schuback and Tora Lane (eds.), *Dis-orientation: Philosophy, Literature and the Lost Grounds of Modernity* (London/Ny: Rowman & Littlefield, 2014), in particular the introduction.

56. Maurice Blanchot. *L'espace littéraire* (Paris: Gallimard, 1955), 91–92, translated by Ann Smock as *The Space of Literature* (Lincoln: University of Nebraska Press, 1982), 77. For a discussion about Maurice Blanchot's views on exile and Jewishness, for his indebtedness to Levinas and to the general question about the figural Jew in French postwar philosophy, see Sarah Mahherschlag, *The Figural Jew: Politics and Identity in Postwar French Thought* (Chicago and London: University of Chicago Press, 2010).

57. Maurice Blanchot, *Le Pas au-delà*, op. cit., 92, *The Step Not Beyond*, op. cit., 65.

58. Ibidem, 8, English, 2.

59. Ibidem.

60. This was a reason for some authors, like Cioran and also in part Beckett, to dislike Blanchot. In his notes on Blanchot's essay on "Lautreamont and Sade," Beckett wrote: "Some excellent ideas, or rather starting-points for ideas, and a fair bit of verbiage, to be read quickly, not as a translator does. What emerges from it though is a truly gigantic Sade, jealous of Satan and of his eternal torments, and confronting nature more than with humankind." *The Letters of Samuel Beckett*, vol. II, 1941–1956 (Cambridge: Cambridge University Press, 2011), 219.

61. Maurice Blanchot, *Le Pas au-delà*, op. cit., 90, *The Step Not Beyond*, op. cit., 63.

62. Ibidem,181, English, 133.

63. Ibidem, 183, English, 134.

64. Ibidem, 88, English, 62.

65. Maurice Blanchot, *L'Écriture du désastre*,178, *The Writing of the Disaster*, 115.

66. Maurice Blanchot, *Le Pas au-delà*, 67, *The Step Not Beyond*, 46.

67. Maurice Blanchot, *L'Écriture du désastre*, 58, *The Writing of the Disaster*, 34.

68. Maurice Blanchot, *Le Pas au-delà*, 72, *The Step Not Beyond*, 50.

69. Maurice Blanchot, *L'Écriture du désastre*, 10, *The Writing of the Disaster*, 3.

70. Maurice Blanchot, *Le Pas au-delà*, 72, *The Step Not Beyond*, 50.

71. Ibidem, 72, English, 50.

72. Ibidem, 81, English, 57.

73. Ibidem.

74. Ibidem, 72, English, 50.

75. Maurice Blanchot, *La part du feu* (Paris: Gallimard, 1949), 327, translated by Charlotte Mandel as *The Work of Fire* (Stanford: Stanford University Press, 1995), 340.

76. Maurice Blanchot, *Le Pas au-delà*, 48, *The Step Not Beyond*, 32.

77. To borrow a formulation by Leslie Hill, in op. cit., 172.

78. Maurice Blanchot, *Le Pas au-delà*, op. cit., 47, *The Step Not Beyond*, op. cit., 31.

79. Ibidem, 48, English, 32.

80. Ibidem, 21, English, 11.

81. Maurice Blanchot, *The Instant of My Death*, op. cit., 3. The question about Blanchot's political commitments is huge and controversial. Fynsk defends strongly the "exilic turn" in Blanchot, where a turn in the very meaning of engagement takes place, and that might have begun already during the period of WWII in conversations with Bataille (see C. Fynsk, op. cit., 3). Michel Surya in his book *L'autre Blanchot: L'écriture de jour, l'écriture de nuit* (Paris: Gallimard, 2015) presents a very different picture, discussing rather Blanchot's omissions, silences, dissimulations, rendering Blanchot's "case" similar to Heidegger's case. My concern here is neither with Blanchot's exilic turn and the redemption of a past to be condemned nor with condemning him for silences and dissimulations. The question here is about Blanchot's thoughts on the imminent temporality of the present tense of the instant, which appears in the mode he presents in this text, which is a mode *between* confession and testimony.

82. Jacques Derrida, *Demeure, L'instant de ma mort*, op. cit., *The Instant of My Death*, op. cit., 43.

83. Ibidem, 52.

84. For an overview of the material for different pictural representations of the death of the emperor Maximiliam, see Juliet Wilson-Bareau, *Manet: The Execution of Maximiliam: Painting, Politics and Censorship* (London: National Gallery Publications, 1992).

85. See Philippe Lacoue-Labarthe, *Agonie terminée, agonie interminable: Sur Maurice Blanchot* (Paris: Galilée, 2011), translated by Hannes Opelz, *Ending and Unending Agony* (New York: Fordham University Press, 2015).

86. Ibidem, 65–66.

87. Ibidem, 81.

88. Maurice Blanchot, *L'espace de la literature*, op. cit., *The Space of Literature*, op. cit.

89. Maurice Blanchot, *Le Pas au-delà*, op. cit, *The Step Not Beyond*, op. cit., 94.

90. Ibidem.

91. Philippe Lacoue-Labarthe, op. cit., 79.

92. Maurice Blanchot, *L'instant de ma mort*, op. cit., *The Instant of My Death*, 49.

93. Derrida, *Demeure, L'instant de ma mort*, op. cit., *The Instant of My Death*, op. cit., 90.

94. Ibidem, 90–91.

95. Maurice Blanchot, *La pas au-delà*, op. cit., *The Step Not Beyond*, op. cit., 72.

96. "Le neutre . . . incline vers la nuit." Ibidem, 104, English, 74.

97. Ibidem, 103.

98. Ibidem, 77.

99. Ibidem, 75.

100. Ibidem, 77.

101. Paul Celan, "The Meridian Speech," in *The Meridian: Collected Prose* (New York: Routledge, 2003), 54.

102. Blanchot's conception of the "arrest" is not without connection with the concept of *nunc stans*, of abiding now, through which the early Christians define eternity "*Nunc fluens facit tempus, nunc stans facit aeternitatum,*" The now that flows away makes time, the now that stands still makes eternity." This quote was attributed to Boethius by Thomas of Aquinas in *Summa Theologica* I, q. 10, art. 2, objection 1, translated by father Laurence Shapcote of the Fathers of the English Dominican province (Chicago: Encyclopædia Britannica; 1990), 41.

103. Ibidem, 76.

104. Marguerite Duras, *Écrire* (Paris: Gallimard, 1993), 53, translated by Mark Polizzotti as *Writing* (Minneapolis: University of Minnesota Press, 2011), 45.

Chapter 4

1. See Emmanuel Catin, *Sérénité. Eckhart, Schelling, Heidegger* (Paris: Vrin, 2012).

2. Martin Heidegger, *Beiträge zur Philosophie*, GA 65 (Frankfurt am Main: Vittorio Klosterman, 1994 [1989]), 388, translated by Richard Roj-cewicz and Daniela Vallega-Neu as *Contributions to Philosophy (Of the Event)* (Bloomington and Indianapolis: Indiana University Press, 2012), 306.

3. Osip Mandelstam, *Tristia*, translated by Joseph Brodsky as *Less Than One: Selected Essays* (London: Penguin 2011 [1986]).

4. For a biography of Clarice, see Benjamin Moser, *Why This World? A Biography of Clarice Lispector* (Oxford: Oxford University Press, 2009).

5. Clarice Lispector, *A paixão Segundo G. H.* (Rio de Janeiro: Editora Rocco, 1998 [1964]), translated by Idra Novey as *The Passion according G. H.* (London: Penguin Books, 2012). Henceforth *Passion*.

6. Ibidem. I had to modify the English translation that reads "and speak therefore of the neutral," when the original says "e falo então em neutro," meaning "and speak therefore in neuter," 169.

7. Hélène Cixous, *Reading with Clarice Lispector (seminar 1980–1985)*, translated by Verena Andermatt Conley (London/Sidney: Harvester Wheatsheaf, 1990), 4.

8. Ibidem, 69.

9. Ibidem, 11.

10. Ibidem, 3.

11. Ibidem, 15.

12. Ibidem, 12.

13. Hélène Cixous, *Readings: The Poetics of Blanchot, Joyce, Kafka, Kleist, Lispector and Tvetayeva* (Minneapolis: University of Minnesota Press, 1991).

14. Ibidem, 1.

15. Ibidem, 14, 23, 30.

16. Ibidem, 3.

17. Clarice Lispcetor, "Literatura de vanguarda no Brasil," in *Outros escritos* (Rio de Janeiro: Rocco, 2005), 98.

18. Nádia Batella Gotlib, *Clarice, uma vida que se conta* (SP: EDUSP; 2013), 480.

19. Clarice Lispector, *Passion*, op. cit., xi. "Este livro é como um livro qualquer. Mas eu ficaria contente se fosse lido apenas por pessoas de alma já formada."

20. Hélène Cixous, *Readings: The poetics of Blanchot, Joyce, Kafka, Kleist, Lispector, and Tsvetayeva*, op. cit., 74.

21. Clarice Lispector, *Água Viva*.

22. Ibidem, 64.

23. Ibidem.

24. Clarice Lispector, "The Fifth Story," in *The Complete Stories*, translated by Katrina Dodson (NY: New Directions Book, 2015), 309–12.

25. Clarice Lispector, *Passion* 5: "Mas por que não me deixo guiar pelo que for acontecendo? Terei que correr o sagrado risco do acaso. E subsitutirei o destino pela probabilidade." 13.

26. Ibidem, 5: "Como é que se explica que o meu maior medo seja exatamente em relação: a ser? E no entanto não há outro caminho. Como se explica que o meu maior medo seja exatamente o de ir vivendo o que for sendo? Como é que se explica que não tolere ver, só porque a vida não é

o que pensava e sim outra - como se antes eu tivesse sabido o que era. Por que é que ver é uma tal desorganização?," 13.

27. 184–85.

28. Ibidem, 60.

29. *Passion*, 123–24.

30. Ibidem, 152.

31. Ibidem.

32. Ibidem, 80.

33. Ibidem, 81.

34. Ibidem, 84.

35. Clarice Lispector, *Perto do Coração Selvagem* (Rio de Janeiro: Rocco, 1998), translated by Alison Entrekin as *Near to the Wild Heart* (NY: New Directions, 2012).

36. Clarice Lispector, *Passion*, 99.

37. Ibidem.

38. Ibidem. "A esperança é um filho ainda não nascido, só prometido, e isso machuca." 147.

39. Ibidem, 69.

40. Ibidem, 75.

41. Ibidem, 14.

42. This expression by James Joyce is the title of Clarice's first published novel, *Perto do coração selvagem*, translated by Alison Entrekin as *Near to the Wild Heart* (NY: New Directions, 2012).

43. Clarice Lispector, *Água Viva* (Rio de Janeiro, editora Artinova, 1973), translated by Stefan Tobler and Benjamin Moser as *Água viva* (NY: New Directions, 2012)

44. Ibidem, 75.

45. Georges Bataille, "Formless," in *Documents 1*, Paris, 1929, 382, translated by Allan Stoekl with Carl R. Lovitt and Donald M. Leslie Jr. as *Georges Bataille, Vision of Excess. Selected Writings, 1927–1939* (Minneapolis: University of Minnesota Press, 1985), 31.

46. For very insightful elucidations of Kant's idea of form, see Rodolphe Gasché, *The Idea of Form: Rethinking Kant's Aesthetics* (Stanford, California: Stanford University Press, 2003), and Juan Manuel Garrido, *La formation des formes* (Paris: Galilée, 2008).

47. Immanuel Kant, *The Cambridge Edition of the Works of Immanuel Kant. Critique of Pure Reason* (Cambridge: Cambridge University Press, 1998).

48. Ibidem.

49. See Michael Marder, "Existential Phenomenology according to Clarice Lispector," in *Philosophy and Literature*, vol. 37 (Baltimore: Johns Hopkins University Press, 2013), 378.

50. Clarice Lispector, *Passion*, 6: "Uma forma contorna o caos, uma forma dá construção à substância amorfa – a visao de uma carne infinita é a visao dos loucos, mas se eu cortar a carne em pedaços e dsitribui-los pelos dias e pelas fomes – então ela não será mais a perdição e a loucura: será de novo a vida humanizada." 14.

51. Sartre describes language as "a sixth finger, a third leg, in short, a pure function which one assimilated," as an extension of the body, see Jean-Paul Sartre. *What Is Literature? And Other Essays* (Cambridge, MA: Harvard University Press, 1988), 35.

52. Clarice Lispector, "Literatura de vanguarda no Brasil" in Outros Escritos (Rio de Janeiro: Rocco, 2005)

53. Ibidem, 98.

54. Ibidem, my translation.

55. Ibidem, 106.

56. Ibidem, 107.

57. Ibidem.

58. Ibidem.

59. Ibidem, 106.

60. Hélène Cixous, *Readings: The Poetics of Blanchot, Joyce, Kafka, Kleist, Lispector, and Tsvetayeva*, op. cit., 1.

61. Hélène Cixous. *Coming to Writing and other Essays* (Cambridge: London: Harvard University Press, 1991).

62. Hélène Cixous, *Readings*, op. cit., 82.

63. Ibidem, 81.

64. "Por que, realmente, como é que se escreve? Que é que se diz? E como dizer? E como é que se começa? E que é que se faz com o papel em branco nos defrontando tranquilo? Sei que a resposta, por mais que intrigue, é única: escrevendo.[. . .] Porque, fora das horas em que escrevo, não sei absolutamente escrever [de *escrita e vida*, p. 25]."

65. *Água Viva*, op. cit., 5.

66. Ibidem.

67. Ibidem.

68. On Hegel's "dash," see Rebecca Comay and Frank Ruda. *Dash— The Other Side of Absolute Knowing* (Cambridge/London: MIT Press, 2018).

69. *Água Viva*, op. cit., 5.

70. Ibidem.

71. Ibidem, 12. "What I say is pure present and this book is a straight line in space."

72. *Passion*, 3: "- - - - estou procurando, estou procurando. Estou tentando entender. Tentando dar a alguém o que vivi e não sei a quem, mas não quero ficar com o que vivi. Não sei o que fazer do que vivi, tenho medo dessa desorganização profunda. Não confio no que me aconteceu." 11.

73. Clarice Lispector, *Passion*,

Qualquer entender meu nunca estará à altura dessa compreensão, pois viver é somente a altura a que possa chegar – meu único nível é viver. Só que agora, agora sei de um segrego. Que já estou esquecendo, ah sinto que já estou esquecendo . . .
         Para sabê-lo de novo, precisaria agora re-morrer. [. . .]. 16

74. Hélène Cixous, "Coming to Writing" *and other Essays* (Cambridge, MA: Harvard University Press, 1991), 67.

75. *Água Viva*, 74.

76. Edmund Husserl, *Collected works*, vol. 4, *On the Phenomenology of the Consciousness of Internal Time (1893–1917)* (The Hague: Nijhoff, 1991).

77. Clarice Lispector, *Passion*, 139. In another passage of *Passion*, we also read: "I had not found a human answer to the enigma. But much more, oh, much more: I had found the enigma itself. I had been given too much," 142.

78. Ibidem.

79. Ibidem.

80. Ibidem.

81. Ibidem, 31.

82. Ibidem, 13–14. "mais um grafismo que uma escrita, pois tento mais uma reprodução do que uma expressão." 21.

83. *Água Viva*, 82.

84. Ibidem, 16.

85. Ibidem.

86. Ibidem.

87. Ibidem, 17.

88. Ibidem.

89. Ibidem.

90. Ibidem, 23.

91. Ibidem, 43.

92. Ibidem.

93. Hélène Cixous, *Readings*, op. cit., 92.

94. *Passion*.

95. *Passion*, 169, translation modified.

96. As mentioned before, Heidegger coined in his "readings" of his own works, the verb *zu isten*, "to is." See note 143.

97. *Água Viva*, 73.

98. Irina Sandomirskaja, "Clarice and Photogeny, or, Not Knowing the Concept of "Enough" in *Ad Marciam* (Stockholm: Södertörn Philosophical Studies, 2017), 201.

99. *Água Viva*, 9.

100. Paixão.

101. Roland Barthes, *Le Neutre: Cours au college de France (1977–1978)* (Paris: Seuil, 2009), 33.

102. *Passion*, 122.

103. Ibidem, 87.

104. Ibidem, 86.

105. Ibidem, 91.

106. Ibidem.

107. Ibidem, 99.

108. Ibidem, 101.

109. Ibidem.

110. See Olga de Sá, *A Escritura de Clarice Lispector* (Petrópolis: Vozes, 1979); Haroldo de Campos, *Metalinguagem & Outras Metas* (São Paulo: Perspectiva, 2013); Benedito Nunes, *O Mundo de Clarice Lispector* (Manaus: Edições do Governo do Estado, 1966); and José Miguel Wisnick, https://vimeo.com/33925444.

111. *Passion*, ibidem, 151, a verdade é o que é, 145.

112. Ibidem, 149.

113. Ibidem, 73.

114. Ibidem, 79.

115. Ibidem, 173.

116. The translation "life just is for me" does not render this strange use of the verbs "to be" and "exist" as transitive verbs. See *Passion*, 189. Clarice accomplishes in sublime beauty what Heidegger tried to do in German when insisting on the need to understand the verb "to be" as a transitive verb. See Martin Heidegger, *Was ist das - die Philosophie?* op. cit., 13.

117. *Passion*, 68 [61].

118. This writing in neuter, in gerundive time—Clarice's passion according to G. H.—maybe could be compared to how Osip Mandelstam's and Paul Celan's poetry is also gerundive and how they praise the gerund (reading Mandelstam's *Journey to Armenia* and Celan's translations and commentaries on Mandelstam's poetry at the German Radio). See Paul Celan, *The Meridian: Final Version—Drafts—Materials* (Stanford: Stanford University Press, 2011).

119. Ossip Mandelstam, "Journey to Armenia," in *The Noise of Time: The Prose of Ossip Mandelstam*, translated by Clarence Brown (San Francisco: North Point Press, 1986), 221–22.

120. *Passion*, 131.

121. Clarice Lispector, *Near to the Wild Heart*, translated by Alison Entrekin (NY: A New Direction Book, 2012), 3.

122. Clarice Lispector, *Near to the Wild Heart*, op. cit., [E a voz, voz de terra. Sem chocar-se com nenhum objecto, maica e longínqua como se tivesse percorrido longos caminhos sob o solo até chegar à garganta, PCS, 74].

123. *Passion*, 186.

A realidade antecede a voz que a procura, mas como a terra antecede a árvore, mas como o mundo antecede o homem, mas como o mar antecede a visão do mar, a vida antecede o amor, a material do corpo antecede o corpo, e por sua vez a linguagem um dia terá antecedido a posse do silêncio.

Eu tenho à medida que designo – e este é o splendor de se ter uma linguagem. Mas eu tenho muito mais à medida que não consigno designar. A realidade é a material-prima, a linguagem é o modo como vou buscá-la- e como não a acho. Mas é do buscar e não achar que nasce o que eu não conhecia, e que instanta-nemante reconheço. A linguagem é o meu esforço humano. Por destino tenho que ir buscar e por destino volto com as mãos vazias. Mas – volto com o indizível. O indizível só me poderá ser dado através do fracasso de minha linguagem. Só quando falha a construção, é que obtenho o que ela não conseguiu. 176

124. Clarice Lispector, *Passion*, 7, "ter a coragem de usar um coração desprotegido e de ir falando para o nada e para o ninguém? Assim como uma criança pensa para o nada. E correr o risco de ser esmagada pelo acaso." 15.

125. Ibidem, 11. "me deixar carente como uma criança que anda sozinha pela terra. Tão carente que só o amor de todo o universe por mim poderia me consolar e cumular." 19.

126. Françoise Van Rossum-Guyon, "A Propos de Manne: Entretien avec Hélène Cixous," in *Hélène Cixous: Chemins d'une écriture* (Vincennes: Presses Universitaires de Vincennes, 1995), 222–23, English version quoted by Susan Rubin Suleiman, "Writing Past the Wall or the Passion according to H. C.," in Hélène Cixous, *Coming to Writing and Other Essays*, op. cit., xx.

127. *Passion*, 157.

128. It was Clarice's sister, Elisa Lispector, who wrote about exile, *No exílio* (RJ: Pongetti, 1945), translated into French *En exil* (Paris: Éditions Des Femmes, 1987).

Chapter 5

1. Marina Tsvetaeva, *Toska po rodine*, "Homesick for the Motherland." This translation of the poem was found on the web, without an author. See https://vdocuments.mx/tsvetaeva-poems.html. Another published translation under the title "Homesickness," which is the elaboration by Elaine Feinstein based on literal versions provided by other translators, can be found in Marina Tsvetaeva, *Selected Poems* (London: Penguin, 1994), 100–101.

2. Sigmund Freud, "Das Unheimliche," first published in *Imago: Zeitschrift für Anwendung der Psychoanalyse auf die Geisteswissenschaften* V (1919), pp. 297–324; *The Uncanny*, translated by David McLintock, introduced by Hugh Haughton (London: Penguin, 2003).

3. Freud, *The Uncanny*.

4. Jean Améry, *At the Minds Limits: Contemplations by a Survivor on Auschwitz and Its Realities*, translated by Sidney Rosenfeld and Stella P. Rosenfeld (Bloomington: Indiana University Press, 1980).

5. Ibidem, 57.

6. Ibidem, 56. Améry uses this expression inspired by the book with same title by Pierre Bertaux, *La mutation humaine* (Paris: Payot, 1964).

7. Ibidem.

8. Herbert Marcuse, *One-Dimensional Man: Studies in the Ideology of Advanced Industrial Society* (Boston: Beacon, 1991 [1964]).

9. Zygmunt Bauman, *Retrotopia* (Malden, MA: Polity, 2017).

10. For Heidegger's discussion about language and Heimat, see "Sprache und Heimat," in *Aus der Erfahrung des Denkens*, GA 13 (Frankfurt am Main: Vittorio Klostermann, 1983), pp. 155–81, and H-G. Gadamer,"Heimat und Sprache," in *Ästhetik und Poetik I. Kunst als Aussage, I Gesammelte Werke*, vol. 8 (Tübingen: Mohr Siebeck, 1993).

11. Martin Heidegger, GA 7, op. cit., English translation by Albert Hofstadter, in *Poetry, Language, Thought* (New York: Harper Colophon Books, 1971).

12. Marina Tsvetaeva, "The Poem of the End," in *Selected Poems*, op. cit., 67.

13. Clarice Lispector, *Água Viva*, op. cit., 23.

14. Ibidem, 50.

15. Marina Tsvetaeva, op. cit.

16. See Hélène Cixous's chapter "Tsvetaeva, Poetry, Passion, and History," in *Readings*, op. cit.

17. Jean-Marie Pontévia, *Tout a peut-être commence par la beauté. Écrits sur l'art et pensées detachées*, op. cit.

18. T. S. Eliot, "Ash-Wednesday," in *Collected Poems, 1909–1962* (New York: Harcourt, Brace & World, 1936), 83 ff.

19. Clarice Lispector, *Água Viva*, op. cit., 68.

20. Ibidem, 9.

21. Ibidem, 87.

22. Ibidem, 66.

Works Cited

Agamben, Giorgio. *La guerre civile. Pour une théorie politique de la stasis* (Paris: Éditions Points, 2015).

Alighieri, Dante. *The Divine Comedy*, translated by Allen Mandelbaum; with an introduction by Eugenio Montale and notes by Peter Armour (New York: Knopf, 1995).

Aquino, Thomas, Laurence Shapcote, and Daniel J. Sullivan. *The summa theologica* (Chicago: Encyclopædia Britannica, 1990).

Arendt, Hannah. *Essays in Understanding, 1930–1954: Formation, Exile and Totalitarianism* (New York: Chock Books, 2005).

Arendt, Hannah. *On Revolution* (New York: Viking, 1963).

Aristotle, *Metaphysics*. Books I–IX (Cambridge, MA: Harvard University Press, 1933).

Aristotle. *On the Soul: Parva Naturalia: On Breath* (Cambridge, MA: Harvard University Press, 1957).

Aristotle. *The Physics*. Books I–IV (Cambridge, MA: Harvard University Press; 1957).

Aristotle. *The Physics*. Books V–VIII (Cambridge, MA: Harvard University Press; 1934).

Ashcroft, Bill, Gareth Griffiths, and Helen Tiffin. *Postcolonial Studies: The Key Concepts* (London: Routledge, 2013).

Augustine. *Confessions*, translated by Henry Chadwick (London: Oxford University Press, 1991).

Badiou, Alain. *Conditions* (Paris: Seuil, 1992).

Badiou, Alain. *Le reveil de l'histoire*, Circonstances 6 (Paris: Nouvelles editions Lignes, 2011).

Balibar, Etienne. *Saeculum, Culture, Religion, Idéologie* (Paris: Galilée, 2012).

Barthes, Roland. *Le Neutre Cours au college de France* (1977–89) (Paris: Seuil, 2009).

Bataille, Georges. *Vision of Excess: Selected Writings, 1927–1939* (Minneapolis: University of Minnesota Press, 1985).

Bauman, Zygmunt. *Retrotopia* (Malden, MA: Polity; 2017).

Beckett, Samuel. *The Letters of Samuel Beckett*, vol. II, 1941–1956 (Cambridge: Cambridge University Press, 2011).

Benvéniste, Emile. *Recherches structurales*, vol. 5, *Publ. à l'occasion du cin-quantenaire de M. Louis Hjelmslev* (Copenhague: Cercle linguistique de Copenhague. Travaux, 1949).

Bertaux, Pierre. *La mutation humaine* (Paris: Payot, 1964).

Blanchot, Maurice *La Communauté inavouable* (Paris: Les Éditions de Minuit, 1983), translated by Pierre Joris, *The Unavowable Community* (Barrytown, NY: Station Hill, 1988).

Blanchot, Maurice. *Faux Pas* (Paris: Gallimard, 1943), translated by Charlotte Mandelas as *Faux Pas* (Stanford: Stanford University Press, 2001).

Blanchot, Maurice. *Le livre à venir* (Paris: Gallimard, 1959), translated by Charlotte Mandel as *The Book to Come* (Stanford: Stanford University Press, 2003).

Blanchot, Maurice. *Pau au-delà* (Paris: Gallimard, 1973), translated by Lycette Nelson as *The Step Not Beyond* (Albany: State University of New York Press, 1992).

Blanchot, Maurice. *L'instant de ma mort* (Paris: Gallimard, 2002 [1994]), translated by Elizabeth Rottenberg as *The Instant of My Death/ Maurice Blanchot*, published together with Jacques Derrida as *Demeure: Fiction and Testimony* (Stanford: Stanford University Press, 2000).

Blanchot, Maurice. *La part du feu* (Paris: Gallimard, 1949), 327, translated by Charlotte Mandel as *The Work of Fire* (Stanford: Stanford University Press, 1995).

Blanchot, Maurice. *L'écriture du désastre* (Paris: Gallimard, 1980) translated by Ann Smock as *The writing of the disaster* (Lincoln: University of Nebraska Press, 1986).

Blanchot, Maurice. *L'espace littéraire* (Paris: Gallimard, 1955), 91–92, tran-slated by Ann Smock as *The Space of Literature* (Lincoln: University of Nebraska Press, 1982).

Blanchot, Maurice, and Thomas l'obscur (Paris: Gallimard, 1950), translated by Lydia Davis, Paul Auster, and Robert Lamberton as "Thomas the Obscure," in *The Station Hill Blanchot Reader*, ed. Goerge Quasha (Barrytown, New York: Station Hill, 1998).

Brodsky, Joseph, *Less Than One: Selected Essays* (London: Penguin, 2011 [1986]).

Campos, Haroldo de. *Metalinguagem & Outras Metas* (São Paulo: Perspectiva, 2013).

Camus, Albert. *L'homme révolté* (Paris: Gallimard,1951).

Catin, Emmanuel. *Sérénité. Eckhart, Schelling, Heidegger* (Paris: Vrin, 2012).

Celan, Paul. *Collected Prose: Paul Celan*. Trans. Rosmarie Waldrop (New York: Routledge, 2007).

Celan, Paul. *Gesammelte Werke in fünf Bänden*, vol. 1 (Frankfurt am Main: Suhrkamp, 1992).

Celan, Paul, *The Meridian: Final Version—Drafts—Materials* (Stanford: Stanford University Press, 2011).

Celan, Paul. "Und mit dem Buch aus Tarussa," in Die Niemandsrose, Fesammelte Werke, vol. 1 (Frankfurt am Main: Suhrkamp, 1992), 288.

Cixous, Hélène. *Chemins d'une écriture* (Vincennes: Presses Universitaires de Vincennes, 1995).

Cixous, Hélène. *"Coming to Writing" and Other Essays* (Cambridge, MA.: Harvard University Press, 1991).

Cixous, Hélène. *Reading with Clarice Lispector*, translated by Verena Andermatt Conley (London/Sidney: Harvester Wheatsheaf, 1990).

Cixous, Hélène. *Readings: The Poetics of Blanchot, Joyce, Kafka, Kleist, Lispector, and Tsvetayeva* (Minneapolis: University of Minnesota Press, 1991).

Comay, Rebecca, and Frank Ruda. *Dash—The Other Side of Absolute Knowing* (Cambridge, MA: MIT Press, 2018).

Dastur, Françoise. *Dire le temps* (La Versanne: Encre marine, 2002), 81.

Derrida, Jacques. "Dialanguages," in *Points . . . Interviews, 1974–1994* (Stanford: Stanford University Press, 1995).

Derrida, Jacques. *Donner la mort* (Paris:Galilée, 1999), translated by David Wills as *The Gift of Death* (Chicago: University of Chicago Press, 1995).

Derrida, Jacques. *D'un ton apocalyptique adopté naguère en philosophie* (Paris: Galilée, 1981), translated by John P. Leavy as "Of an Apocalyptic Tone Recently Adopted in Philosophy, *The Oxford Literary Review*, vol. 6, no. 2 (1984).

Derrida, Jacques. *Edmund Husserl, L'origine de la géométrie* (Paris: PUF, 1962), translated by John P. Leavey Jr. *as Edmund Husserl's Origin of Geometry: An introduction* (Lincoln: University of Nebraska Press, 1989 [1978]).

Derrida, Jacques. *Heidegger: La question de l'être et de l'histoire. Cours de l'ENS-Ulm (1964–65)* (Paris: Galilée, 2013), translated by Geoffrey Bennington as *Heidegger: The Question of Being and History* (Chicago: University of Chicago Press, 2016).

Derrida, Jacques. *Le monologinguisme de l'autre ou La Protèse d'origine* (Paris: Galilée, 1996), translated by Patrick Mensah as *Monolinguism of the Other or the Prosthesis of Origin* (Stanford: Stanford University Press, 1998).

Derrida, Jacques. "No Apocalypse, Not Now (Full Speed Ahead, Seven Missiles, Seven Missives," in *Diacritics* 14, no. 2 (Summer, 1984).

Derrida, Jacques. *Spectres de Marx, the State of the Debt, the Work of Mourning, and the New International* (New York: Routledge, 1994).

Derrida, Jacques. *Le toucher, Jean Luc Nancy* (Paris: Galilée, 2000), translated by Christine Irizarry as *On Touching* (Stanford: Stanford University Press, 2005).

Di Cesare, Donatella. "Exile: Human Condition in the Globalized World," in *Philosophy Today*, SPEP Supplement, 2008.

Dickinson, Emily. *The Complete Poems of Emily Dickinson*, edited by Thomas H. Johnson (Boston/Toronto: Little, Brown and Company, 1960).

Diderot, *Élements de phyisologie* (Paris: Didier, 1964).

Duras, Marguerite. *Écrire* (Paris: Gallimard, 1993), translated by Mark Polizzotti, *Writing* (Minneapolis: University of Minnesota Press, 2011).

Eliot, T. S. *Collected Poems, 1909–1962* (New York: Harcourt, Brace & World, 1936).

Eliot, T. S. *Selected Prose* (San Diego: Harcourt Brace & Company, 1975).

Farocki, Harun. *Aufschub. Respite.* https://www.youtube.com/watch?v=2ciiA0by SPc.

Flusser, Vilém. Pós-historia: *Vinte instantâeos e um modo de usar* (SP: Annablume, 2011), translated by Rodrigo Maltez Novaes as *Post-History* (University of Minnesota Press, 2013).

Foucault, Michel. *Les mots et les choses: une archéologie des sciences humaines* (Paris: Gallimard, 1995), translated as *The Order of Things: An Archaeology of the Human Sciences* (London: Repr., Tavistock/Routledge, 1989 [1974]).

Foucault, Michel. *The Thought from Outside* and Maurice Blanchot, *Michel Foucault as I Imagine Him* (New York: Zone Books, 1990).

Freud, Sigmund. *The Uncanny*, translated by David McLintock; introduced by Hugh Haughton (London: Penguin, 2003).

Fynsk, Christopher. *Last Steps: Maurice Blanchot's Exilic Writing* (New York: Fordham University Press, 2013).

Gadamer, Hans-Georg. *Ästhetik und Poetik I. Kunst als Aussage, I Gesammelte Werke*, vol. VIII (Tübingen: Mohr Siebeck, 1993).

Gadamer, Hans-Georg. *Gadamer Lesebuch* (Tübingen: Mohr Siebeck, 1997).

Garrido, Juan Manuel. *La formation des formes* (Paris: Galilée, 2008).

Gasché, Rodolphe. *The Idea of Form: Rethinking Kant's Aesthetics* (Stanford: Stanford University Press, 2003).

Gehlen, Arnold. *Studien zur Anthropologie und Soziologie* (Neuwid, Luchterhand, 1963).

Goethe, Johan Wolfgang, and Gordon L. Miller. *Metamorphosis of Plants* (Cambridge, MA: MIT Press, 2009).

Gotlib, Nádia Batella, *Clarice, uma vida que se conta* (SP: EDUSP, 2013).

Groys, Boris. *Osteuropäische Kulturen im Zeitalter des Postkommunismus* (Frankfurt am Main: Suhrkamp, 2005).

Hägglund, Martin. *Radical Atheism: Derrida and the Time of Life* (Stanford: Stanford University Press, 2008).

Halbwachs, Maurice. *La mémoire collective* (Paris: PUF, 1950), translated by Lewis A. Coser as *On Collective Memory* (Chicago: Chicago University Press, 1992).

Hamacher, Werner. "Afformative, Strike," 13 *Cardoso Law Review* (1991).

Hamacher, Werner. "Geschichte literarischer und phänomenaler Ereignisse," in Marja Rauch and Achim Geisenhanslücke (eds.) *Texte zur Theorie und Didaktik der Literaturgeschichte* (Stuttgart: Reclam, 2012), 169.

Handke, Peter. *Till Day You Do Part or a Question of Light*, translated by Mike Mitchell (London: Seagull Books, 2010).

Heidegger, Martin. *Aus der Erfahrung des Denkens*, Gesamtausgabe, vol. 13 (Frankfurt am Main: Vittorio Klostermann, 1983).

Heidegger, Martin. *Beiträge zur Philosophie* GA 65 (Frankfurt am Main: Vittorio Klostermann, 3 ed. 2003) translated by Richard Rojcewicz and Daniela Vallega-Neu (Indiana: Indiana University Press, 2012).

Heidegger, Martin. *Bremer und Freiburger Vorträge* GA 79 (Frankfurt am Main: Vittorio Klostermann, 1994), translated by Andrew Mitchell as *Bremen and Freiburg Lectures: Insight into* That Which Is *and* Basic Principles of Thinking (Bloomington: Indiana University Press, 2012).

Heidegger, Martin. "Die Frage nach der Technik," in *Vorträge und Aufsätze* GA 07 (Frankfurt am Main: Vittorio Klostermann, 2000), translated by William Lovitt as *The Question Concerning Technology and Other Essays* (New York: Harper Torchbooks, 1977.

Heidegger, Martin. Die *Grundbegriffe der Metaphysik. Welt—Endlichkeit—Einsamkeit*, GA 29–30 (Frankfurt am Main:Vittorio Klostermann,1983), translated by William McNeill and Nicholas Walker as *The Fundamental Concepts of Metaphysics: World, Finitude, Solitude* (Bloomington and Indianapolis: Indiana University Press, 1995).

Heidegger, Martin. *Identität und Differenz* (Pfullingen: Günther Neske, 1957).

Heidegger, Martin,. *Nietzsche II*, GA 6.2, translated by David Farell Krell, Nietzsche (San Francisco: Harper & Row, 1979).

Heidegger, Martin. *Poetry, Language, Thought* (New York: Harper Colophon Books, 1971).

Heidegger, Martin. *Sein und Zeit* GA 2 (Frankfurt am Main: Vittorio Klostermann, 1977), 170, translated by Joan Stambaugh as *Being and Time* (NY: State University of New York Press, 1996).

Heidegger, Martin. *Vorträge und Aufsätze*, GA 7 (Frankfurt am Main: Vittorio Klostermann, 2000), translated by David Farell Krellas as *Basic Writings* (London: Taylor & Francis, 2010).

Heidegger, Martin. *Was heißt Denken?* GA 8 (Frankfurt am Main: Vittorio Klostermann, 2002), translated by G. Genn Gray as *What Is Called Thinking?* (New York: HarperCollings, 1976).

Heidegger, Martin. *Was ist das—die Philosophie?* (Pfullingen: Günther Neske, 1956).

Heidegger, Martin. *Wegmarken* GA 9 (Frankfurt am Main: Vittorio Klostermann, 1978), edited by William Mc Neill as *Pathmarks* (Cambridge: Cambridge University Press, 1998).

Heidegger, Martin. *Zur Sache des Denkens* GA 14 (Frankfurt am Main: Vittorio Klostermann, 1962–64), translated by Joan Stambaugh as *Time and Being* (New York: Harper & Row, 1972).

Hill, Leslie. *Maurice Blanchot and fragmentary Writing: A Change of Epoch* (London/New York: Continuum, 2012).

Hobbes, Thomas. *Leviathan* (New York: Norton, 1997).

Hölderlin, Friedrich. *Sämtliche Werke*. Grosse Stuttgarter Ausgabe (Stuttgart: Kohlhammer/Cotta Verlag, 1961).

Horton-Smith, L. "The Origin of the Gerund and Gerundive," in *The American Journal of Philology* 15, no. 2 (1894).

Husserl, Edmund. *Cartesian Meditations. An Introduction to Philosophy*, translated by Dorion Cairns (The Hague: Martinus Nijhoff, 1960).

Husserl, Edmund. *Collected Works*. Vol. 4, *On the Phenomenology of the Consciousness of Internal Time (1893–1917)* (The Hague: Nijhoff, 1991).

Husserl, Edmund. *Husserliana: Gesammelte Werke. Bd 29, Die Krisis der europäischen Wissenschaften und die transzendentale Phänomenologie, Ergänzungsband, Texte aus dem Nachlass 1934–1937* (Dordrecht: Kluwer; 1993), translated by David Carr as *The Crisis of European Sciences and Transcendental Phenomenology: An Introduction to Phenomenological Philosophy* (Evanston: Northwestern University Press, 1978 [1970]).

Husserl, Edmund. *Phantasie, Bildbewusstsein, Erinnerung. Zur Phänomenologie der anschaulichen Vergegenwärtigungen. Texte aus dem* Nachlass (1898–1925). Husserliana bd 23 (The Hague: Nijhoff, 1980); English translation *Phantasy, Image Consciousness, and Memory (1898–1925)*, Collected Works, vol. 11 (Dordrecht: Springer, 2005).

Jaeger, Werner. *Paideia: Die Formung des griechischen Menschen.* [1:] 4. Aufl. [2–3:] 3. Aufl., Berlin, 1959, translated as *Paideia: The Ideals of Greek Culture* (New York: Oxford University Press, 1939–1944).

Jameson, Fredric. *The Seeds of Time* (New York: Columbia University Press, 1994).

Kant, Immanuel. *The Cambridge Edition of the Works of Immanuel Kant: Critique of Pure Reason* (Cambridge: Cambridge University Press, 1998).

Kant, Immanuel. *Political Writings* (Cambridge: Cambridge University Press, 1991).

Kant, Immanuel. *Religion and Rational Theology* (Cambridge: Cambridge University Press, 1996).

Kant, Immanuel. *Theoretical Philosophy after 1781* (Cambridge: Cambridge University Press, 2002).

Kant, Immanuel. *Werke*. Akademie-Textausgabe (Berlin: Gruyter, 1968).

Kosuth, Joseph. *Art after Philosophy and After: Collected Writings, 1966–1990* (Cambridge, MA: MIT Press, 1991).

Krell, David Farrell. *Ecstasy, Catastrophe: Heidegger from* Being and Time *and the* Black Notebooks (Albany: State University of New York Press, 2015).

Lacoue-Labarthe, Philippe. *Agonie terminée, agonie interminable sur Maurice Blanchot: Suivi de L'émoi* (Paris: Galilée, 20119), translated by Hannes Opelz as *Ending and Unending Agony. On Maurice Blanchot* (New York, Fordham University Press, 2015).

Lacoue-Labarthe, Philippe. *Typographie* (Paris: Aubier-Flammarion, 1975), edited by Christopher Fynsk as *Typography* (Stanford: Stanford University Press, 1989).

Lane, Tora, and Marcia Sá Cavalcante Schuback (eds.). *Dis-orientations: Philosophy, Literature and the Lost Grounds of Modernity* (London: Rowman & Littlefield, 2015).

Langlois, Christopher. "Temporal Exile in the Time of Fiction: Reading Derrida Readig Blanchot's *The Instant of my Death*," in *Mosaic* 48, no. 4 (2015): 17–32. Special Issue: *A Matter of Life/Death*.

Levinas, Emmanuel. *Autrement qu'être ou au-delà de l'essence* (Netherlands: Springer, 1974), *Otherwise Than Being, or Beyond Essence* (Pittsburgh, PA: Duquesne University Press, 1998–[1974]).

Levinas, Emmanuel. *De l'existence à l'existant* (Paris: Vrin, 2. éd. augm., 1986), translated by Alphonso Lingis as Existence and existents (Pittsburgh, PA: Duquesne University Press, 2001 [1978]).

Levinas, Emmanuel. "L'existentialisme, l'angoisse et la mort," in *Exercices de la patience. Heidegger*, nos. 3–4 (Paris: Obsidiane, 1982).

Levinas, Emmanuel. "Reality and Its Shadow," in *Unforseen History* (Urbana: University of Illinois Press, 2004).

Lispector, Clarice. *Água Viva* (Rio de Janeiro, editora Artinova, 1973), translated by Tobler, Stefan, and Moser, Benjamin as *Água Viva* (New York: New Directions, 2012).

Lispector, Clarice. *The Complete Stories*, translated by Katrina Dodson (New York: New Directions Book, 2015).

Lispector, Clarice. *Outros escritos* (Rio de Janeiro: Rocco, 2005), translated by Alison Entrekin as *Near to the Wild Heart* (New York: New Directions Book, 2012).

Lispector, Clarice. *A paixão Segundo G. H.* (Rio de Janeiro: Editora Rocco, 1998).

Lispector, Clarice. *The Passion according to G. H.* (London: Penguin Books, 2012).

Lispector, Clarice. *Perto do Coração Selvagem* (Rio de Janeiro: Rocco, 1998).

Loraux, Nicole. *La cite divisée* (Paris: Payot, 1997), translated as *The Divided City: On Memory and Forgetting in Ancient Athens* (New York. Zone Books, 2002).

Löwith, Karl. *The Meaning in History* (Chicago: University of Chicago Press, 1949).

Lyotard, Jean-François. *La condition postmoderne: rapport sur le savoir* (Paris: Éd de minuit, 1979), English translation *The Postmodern Condition: A Report on Knowledge* (Manchester: Manchester University Press, 1984).

Mahherschlag, Sarah *The Figural Jew: Politics and Identity in Postwar French Thought* (Chicago: University of Chicago Press, 2010).

Malabou, Catherine. *Plasticité* (Paris: Éditions Léo Scheer, 1999), *L'Avenir de Hegel: Plasticité, Temporalité, Dialectique* (Paris: Vrin, 1996), translated by Lisabeth During as *The Future of Hegel: Plasticity, Temporality, and Dialectic* (New York: Routledge, 2004).

Maldiney, Henri. *Aîtres de la langue et demeures de la pensée* (Lausanne: L'Âge d'Homme, 1975).

Mandelstam, Osip. *The Noise of Time: The Prose of Osip Mandelstam*, translated by Clarence Brown (San Francisco: North Point, 1986).

Marcuse, Herbert *One-dimensional Man: Studies in The Ideology of Advanced Industrial Society* (Boston: Beacon, 1991 [1964]).

Marder, Michael. "Existential Phenomenology according to Clarice Lispector," in *Philosophy and Literature*, vol. 37 (Baltimore: John Hopkins University Press, 2013).

Marder, Michael. *Groundless Existence: The Political Ontology of Carl Schmitt* (New York: Continuum, 2010).

Minaudier, Jean-Pierre. *Poésie du Gérondif (vagabondages linguistiques d'un passion de peuples et de mots)* Le Tripode, 2014.

Mitchell, Andrew J. *The Fourfold: Reading the Late Heidegger* (Evaston, IL: Northwestern University Press, 2015).

More, Thomas. *Libellus vere aureus, nec minus salutaris quam festivus, de optimo rei publicae statu deque nova insula Utopia*, English translation *A Fruitful and Pleasant Work of the Best State of a Public Weal, and of the end of the New Isle Called Utopia*, 1516. http://www.ub.uni-bielefeld.de/diglib/more/utopia/.

Moser, Banjamin. *Why This World? A Biography of Clarice Lispector* (Oxford: Oxford University Press, 2009).

Nabokov, Vladimir. *Speak, Memory: An Autobiography Revisited* (London: Penguin Books, 1967).

Nancy, Jean-Luc. *The Birth to Presence* (Stanford: Stanford University Press, 1993).

Nancy, Jean-Luc. *La communauté désouvrée* (Paris: Christian Bourgois Éditeur, 1999).

Nancy, Jean-Luc. *L'Équivalence des catastrophes* (Après Fukushima) (Paris: Galilée, 2012).

Nancy, Jean-Luc. "L'existence exile," in *Cahiers Intersignes*, edited by Fethi Benslama, 14–15 (2001).

Nancy, Jean-Luc. *Que faire?* (Paris: Galilée, 2016).

Nancy, Jean-Luc, and Marcia Sá Cavalcante Schuback (eds.). *Philosophy Today*, special issue *History, Today* (Chicago: De Paul University, 2016).

Nietzsche, Friedrich. *Die fröhliche Wissenschaft*, no. 342, KSA, 23, dtv/de Gruyter.

Nietzsche, Friedrich. *Nachgelassene Fragmente 1885–1887*, KSA, 2232, 104.

Novalis. *Schriften: die Werke Friedrich von Hardenbergs*. Bd 3, *Das philosophische Werk* 2 (Stuttgart: Kohlhammer,1983).

Nunes, Benedito. *O Mundo de Clarice Lispector* (Manaus: Edições do Governo do Estado, 1966).

Patočka, Jan. *Heretical Essays in the Philosophy of History* (Chicago: Open Court, 1996).

Pessoa, Fernando. *Obra poética* (Rio de Janeiro: Nova Aguilar, 1983).

Plutarch. *Moralia*, vol. VII (Harvard: Loeb Classical Library edition, 1959).

Pontévia, Jean-Marie. *Tout a peut-être commence par la beauté. Écrits sur l'art et pensées detachées* (William Blake & Co. Édit. 1985).

Proust, Françoise. *L'Histoire à contretemps. Le temps historique chez Walter Benjamin* (Paris: Editions du Cerf, 1994).

Pseudo Dionysius, *The Complete Works* (New York: Paulist, 1987).

Quignard, Pascal. *Sur l'image qui manque à nos jours* (Paris: Arléa, 2014).

Rancière, Jacques. *En quel temps vivons-nous?* (Paris: La Fabrique Éditions, 2017).

Rat, Ramona. *Un-common Sociality: Thinking Sociality with Levinas* (Stockhlm: Södertörn Philosophical Studies 19, 2016).

Rauch, Marja, and Achim Geisenhanslücke (eds.). *Texte zur Theorie und Didaktik der Literaturgeschichte* (Stuttgart: Reclam, 2012).

Ricci, Felicetti. "Gagner la Heimatlosigkeit" in *Heidegger Studies*, vol. 24 (Berlin: Duncker & Humblot, 2008).

Richter, Gerhard. *Afterness: Figures of Following in Modern Thought and Aesthetics* (New York: Columbia University Press, 2011).

Ricoeur, Paul. *La mémoire, l'histoire, l'oubli* (Paris: Seuil; 2000), *Memory, History, Forgetting* (Chicago: University of Chicago Press, 2004).

Robinson, Marc (ed.). *Altogether Elsewhere: Writers on Exile* (Winchester, MA: Faber and Faber, 1994).

Romano, Claude. *L'événement et le monde* (Paris: Puf, 1999). English version *Event and World* (New York: Fordham University Press, 2009).

Rosenzweig, Franz. *The Star of Redemption* (Madison: University of Wisconsin Press, 2005).

Sá, Olga de. *A Escritura de Clarice Lispector* (Petrópolis: Vozes, 1979).

Said, Edward. *Reflections on Exile, & Other Literary & Cultural Essays* (London: Granta, 2000).

Sallis, John. *Force of Imagination: The Sense of the Elemental* (Bloomington: Indiana University Press, 2000).

Sandomirskaja, Irina. "Clarice and Photogeny, or, Not Knowing the Concept of 'Enough,'" in Hans Ruin and Jonna Bornemark (eds.), *Ad Marciam* (Stockholm: Södertörn Philosophical Studies, 2017).

Sartre, Jean-Paul. *What Is Literature? And Other Essays* (Cambridge, MA: Harvard University Press, 1988).

Schelling, Friedrich. *Initia Philosophiae Universea: Erlanger Vorlesung WS 1820/21* (Bonn: Bouvier, 1969).

Schelling, J. W. *System des transzendentalen Idealismus* (Hamburg: Felix Meiner, 1957), *System of transcendental Idealism* (Charlottesville: University Press of Virginia, 1978).

Schuback, Marcia Sá Cavalcante, and Jean-Luc Nancy (eds). *Being with the Without* (Stockholm: Axl Books, 2013).

Schuback, Marcia Sá Cavalcante, and Susanna Lindberg (eds.). *The End of the World: Contemporary Philosophy and Art* (London/Ny: Rowman & Littlefield, 2017).

Seneca. *Moral Essays, Volume II: De Consolatione ad Marciam. De Vita Beata. De Otio. De Tranquillitate Animi. De Brevitate Vitae. De Consolatione ad Polybium. De Consolatione ad Helviam.* Translated by John W. Basore. Loeb Classical Library 254 (Cambridge, MA: Harvard University Press, 1932).

Sheehan, Thomas. *Heidegger: The Man and the Thinker* (Chicago: Precedent, 1981).

Spivak, Gayatri Chakravorty, *A Critique of Postcolonial Reason: Toward a History of the Vanishing Present* (Cambridge, MA: Harvard University Press, 1999).

Surya, Michel. *L'autre Blanchot. L'écriture de jour, l'écriture de nuit* (Paris: Gallimard, 2015).

Szondi, Peter. "Reading 'Engführung': An Essay on the Poetry of Paul Celan," in *Boundary 2* 11, no. 3, *The Criticism of Peter Szondi* (Spring, 1983): 231–64 (Duke University Press).

Trawny, P., and A. Mitchell. *Heidegger, die Juden, nocheinmal* (Frankfurt am Main: Vittorio Klostermann, 2015).

Trawny, Peter. *Freedom to Fail: Heidegger's Anarchy* (New Jersey: John Wiley & Sons, 2015).

Trawny, Peter. *Heidegger and the Myth of a Jewish World Conspiracy* (Chicago: University of Chicago Press, 2015).

Tsvetaeva, Marina. *Selected Poems* (London: Penguin, 1994).

Valéry, Paul. *Oeuvres I* (Paris: Gallimard, 1957).

Vallega, Alejandro. *Heidegger and the Issue of the Space: Thinking on Exilic Grounds* (Philadelphia: Pennsylvania State University Press, 2003).

Van Hulle, Dirk. "Sans." *The Literary Encyclopedia*. First published March 1, 2004.

Varro, Marcus Terentius. *On the Latin Language*. [2], Books VIII–X Fragments, Rev. (Cambridge, MA: Harvard University Press, 1951).

Wackernagel, Jakob. *Lectures on Syntax: With Special Reference to Greek, Latin, and Germanic* (Oxford: Oxford University Press, 2009).

Weber, Samuel. *Theatricality as Medium* (New York: Fordham University Press, 2004).

Wilson-Bareau, Juliet. *Manet: The Execution of Maximilian. Painting, Politics and Censorship* (London: National Gallery Publications, 1992).

Wisnick, José Miguel. https://vimeo.com/33925444.

Wolfe, Thomas. *You Can't Go Home Again*. With an introduction by Edward C. Aswell (New York, 1941).

Wood, David C. *Time after Time* (Bloomington: Indiana University Press, 2007).

Wybrands, F. "La rencontre, l'excès: Heidegger—Blanchot," in *Exercices de la patience. Blanchot* (Paris: Obsidiane, 1981), 79–87.

Young, Robert J. C. *Postcolonialism: An Historical Introduction* (Oxford: Blackwell, 2001).

Zarader, Marlène. *L'être et le neutre. À partir de Maurice Blanchot* (Lagrasse, editions Verdier, 2001).

Ziarek, Krzysztof. *Inflected Language: Tomwards a Hermeneutics of Nearness: Heidegger, Levinas, Stevens, Celan* (Albany: State University of New York Press, 1994).

Zizek, Slavov. *Living in the End Times* (London/NY: Verso, 2010, 2011).

Index

afterness of existence
  postexistential condition and, 20–21
  questioning past-future
    connections, 19–20
  uncanny movement of present,
    22
Água Viva (Lispector) (novel)
  gerundive mode of time and, 12
  gift of attention to each thing,
    124
  how to read, 85–86
  improvisation connected with act
    of seeing, 103–5
  writing style explained in, 99
  See also *Passion according to
    G. H., The*
Alzheimer's disease, 27
Améry, Jean, loss of home and,
  117–18
"approaching reading," defined,
  12–13
Arendt, Hannah, 8
Aristotle
  Heidegger calling out mistake of,
    52
  on memory, 23, 24
  Memory and Recollection, 24, 52
  Metaphysics, 7
  movement as ecstatic, 19
  Physics, 18
  on time and change, 36
  "unmoved mover" concept and,
    18
Ash-Wednesday (Eliot), 125

Barthes, Roland, 107
Bataille, Georges, 61, 93
Bauman, Zygmunt, 119
Benjamin, Walter, 7, 21, 25
Benveniste, Emile, 36
Black Notebooks (Heidegger), 41,
  44, 122
Blanchot, Maurice
  overview, 12, 61–62
  approach to is-being and, 83–85
  compared to Clarice Lispector, 98
  Critique, 74
  Death Sentence (novel), 78–79
  exilic existence viewed by, 63–64,
    66
  False Steps (essays), 67
  figure of the between and, 64
  Foucault on, 69
  fragmentary writing and, 62–67
  Fynsk on, 62–63
  Heidegger and, 61
  "Heraclitan" style and, 60
  on "immobile becoming," 22
  Infinite Conversation, The, 62, 63,
    64

Blanchot, Maurice (continued)
  influence on French thought and
    literature, 59–60
  Instant of My Death, The, 67,
    77–79
  interconnection with Heidegger
    and Lispector, 9–11
  Kafka and, 71–72
  Levinas and, 70–71
  "Literature and the Right to
    Death" (essay), 69, 74
  motif of "the step" and, 67
  the neuter in, 107–8
  not author of exile, 62
  not-departing and, 8
  presence and, 74, 120
  sovereignty of the nothing and,
    73–74
  stepping beyond obsession of,
    60–61
  tension of the imminent and,
    66–67
  theatrical fictions of, 67
  Thomas the Obscure, 67
  on thought of withdrawal, 12, 60
  writing concern of, 60, 68–69,
    72–73
  See also Derrida, Jacques; Step
    not Beyond, The; Writing of the
    Disaster, The
Book of Disquiet (Pessoa), 8
book's challenge, 29–30
book's structure, 11–12
book's thesis, 7

capitalism, 15, 17–18, 118–19
care, defined, 39–40
carpe diem, 123
Celan, Paul, 77, 81, 126
Char, René, 63
Chernyshevsky, Nikolai, 16

Chronicles for Young People
    (Lispector), 99
Cixous, Hélène, 85–86, 87, 98, 102
Contributions to Philosophy
    (Heidegger), 33, 41, 44, 46, 84
countertime, defined, 4
Crisis of European Sciences (Husserl),
    17
Critique (Blanchot), 74

Dasein
  difficulty in translating, 33
  as "existing," 34
  "neutrality" of, 50–51
  as out of itself, 39–40
  "turn" (Kehre) in, 43
Dastur, Françoise, 47
Death Sentence (Blanchot) (novel),
    78–79
Derrida, Jacques
  Being and Time seminar and, 34
  Blanchot and, 59–60, 77–78
  on "memory of the other," 26–27
  Step [not]Beyond and, 79–80
  on thinking as deconstruction,
    23, 27
  See also Blanchot, Maurice
Duras, Marguerite, 82

echography, 109
ecstasy of time (Heidegger
    conversation), 31–54
ecstatic temporality, 32
Eliot, T. S., Ash-Wednesday, 125
eternal detour, defined, 70–71, 84
excesses of today
  anxiety and, 16–17
  Aristotle and "unmoved mover,"
    18
  ecstasy and, 18–19
  emptiness of omnipresence, 17

excess defined, 16
"at the flower of the skin," 17
as "our" time, 16
reversal of movement *inside*
    movement, 18
times of excess or exile? 16–17
un-transcendental excess, 17
without transcendence, 17–18
See also capitalism
exile
    defined ontologically, 1–2
    as present tense verb, 120–22
exilic detour, 72, 81
exilic existence, 32, 66
exilic fragmentation, 63–64
existence in exile, 7–8, 32

False Steps (Blanchot) (essays), 67
"Fifth Story, The" (Lispector) (short
    story), 89–90
"Formless" (Bataille), 93
Foucault, Michel, on Blanchot, 69
fragmentary writing, 62–67
Freud, Sigmund
    Das Unheimliche, The Uncanny,
    116–17
    on memory, 26
from-to, 4, 7, 18–19
future, an absence without traces,
    24–25
Fynsk, Christopher, on Blanchot,
    62–63

Gadamer, Hans-Georg, concept of
    Verweilung and, 22
Gauguin, Paul, 20
Gelassenheit, 83
geometry, origin of, 58
"gerund"
    defined, 5–7
    existence and, 8

gerundive time, 8, 67–68, 82,
    98–105, 108–11
globalism, 15, 119

Hamacher, Werner, 4
Hegel, Georg
    figure of the negation of negation
    and, 74
    Phenomenology of the Spirit, 2, 100
    Science of Logic, 100
Heidegger, Martin
    overview, 12
    "approaching reading" of thoughts
    of, 53–54
    approach to is-being, 83–85
    on Aristotle's mistake of seizing
    present, 52
    being as being in an
    aletheiological way, 45
    Black Notebooks, 41, 44, 122
    "call of consciousness," 39
    care and, 39–40
    to-come as *overcoming*, 41
    Contributions to Philosophy, 33, 41,
    44, 46, 84
    Dasein, 33, 34, 39–40, 43
    destiny and, 41
    ecstasy of time, 12
    ecstatic temporality, 32–33, 37,
    38
    end-of-world obsession with (30s
    and 40s), 45–46
    epochal changes of meaning of
    being, 51–52
    "exile" absent from work, 31–32,
    54
    existence, 34–39, 53–54
    forgetfulness of being and, 41–42
    formula *Es gilt in Time and Being*,
    52
    on fugitive philosopher, 58–59

Heidegger, Martin (continued)
grammar and, 42
home and homeland, 122
hyphenation, 33, 34
interconnection with Blanchot
and Lispector, 9–11
on intimacy and insistency of
being, 57
Kant and, 42–43
"leap into the event of presence,"
47
Levinas and, 40
lightning example and being
(Nietzsche), 44
listening and thinking anew,
49–50
Metaphysical Foundations of
Logic (1928), 50
"neutrality" of Dasein., 50–51
old words used by, 47
ontological narrative and
modernity, 43
overcoming metaphysics, 55–56
performativity of thinking-writing
and, 48
philosopher of the excess, 31–32
positionality of being, 48
"Question Concerning
Technology" (lecture), 48
question of being and, 42–43, 44
rethinking of Being and Time, 49
rootedness and, 8
Time and Being (lecture), 47–49
transparency (Durchsichtigkeit)
and, 38
truth and, 59
"The Turn" (Die Kehre), 47
vocabulary of existence, 34–35
What Is Called Thinking?
(Heidegger) (lecture series
1952), 57–58

What Is That Philosophy? 52–53
while and meanwhile as focus, 47
while-withdrawing of being
described, 46
writing difference and thinking, 58
Hill, Leslie, 62–63, 65
Hobbes, Thomas, on memory, 25
Hölderlin, Friedrich, 44, 48
home
globalism and backward longing,
15, 119
learning to inhabit, 122–23
Tsvetaeva on, 123
the uncanny of (Freud), 116–17,
119
homelessness without end (exile),
114
See also Homesick for the
Motherland
Homer, Odyssey, 2
Homesick for the Motherland
(Tsvetaeva) (poem text),
114–16
human soul, Seneca on, 2
Husserl, Edmund, 58
Crisis of European Sciences, 17
critique of modern objectification
and subjectivism, 43
flux of conscious of internal time
and, 102
on memory, 24–25

improvisation, 103–5
Infinite Conversation, The
(Blanchot), 62, 63, 64
Insight into that which is (series)
(Heidegger), 47
Instant of My Death, The
(Blanchot), 67, 77–79
intimacy and insistency, Heidegger
and, 57

is-being, dwelling in the
density and weight of is-being
and, 120
Eliot's strategy of listening, 125
exile as present tense verb,
120–22
existing *with a without*, 123
Heidegger's thoughts on home
and homeland, 122
homelessness without end (exile),
114
lesson in dwelling in the
gerundive, 123–24
as "shivering bird in the hand,"
8, 105, 113
struggle for presence, 74, 120
tensioned between marks called
existence, 119–20
See also globalism and backward
longing; home; *Homesick for the
Motherland*; Uncanny, The
is-being, literature of the, 106–11
continuation of interruption, 106
"the neutral crafting of life," 108
writing in neuter is writing in
gerundive time, 108–11
is-existing, gerund of being, 7–8,
13, 29, 121, 126

Jewish tradition, 2–3
Joyce, James, 87

k, 90–91
Kafka, Franz
cockroach and, 89
Kafkaesque readings of Clarice,
85–86
Space of Literature, The, 71–72,
79
Kant, Immanuel
critical philosophy of, 3, 93–95

eye-I-centered schema, 37
interpretation of Descartes, 69
Klee, Paul, 59

Lacoue-Labarthe, Philippe, 18,
78–79
leaving, experience of, 84
Levinas, Emmanuel
Blanchot and, 71–72, 83
on Heidegger's hyphenation, 33,
34
"Reality and its shadow" (essay),
69
understanding of exile and, 70
Wahl and, 40
Lispector, Clarice
overview, 12–13
Brazilian writer of renown, 85
Chronicles for Young People, 99
compared to Blanchot, 98
critical about current notions
of avant-garde as formal
innovation, 96–97
dash used by, 99–100
exile rarely spoken of by, 111
"The Fifth Story," 88–90
fragments and, 102–3
gift of attention to each thing,
124
improvisation and, 103–5
on the "infinite monstrous meat,"
of passion, 94–95
interconnection with Heidegger
and Blanchot, 9–11
is-being, literature of the, 107–8
in the *is* of each thing, 123, 124
it is-being: or the neuter crafting
of life, 106–12
on language as way of searching,
110
lecture on literature by, 96

Lispector, Clarice (*continued*)
  Near to the Wild Heart, 86–87,
    109
  quick reading of works expected
    by, 99, 100
  on reading her works by heart,
    102
  reading time and time of reading,
    83–97
  risk of writing in gerundive time,
    98–105
  speaking within the is being,
    125–26
  speaks and writes in neuter,
    106–7
  strong role of gerund in language
    of, 98
  unique style of, 85–89, 99–100,
    103, 108–9, 126
  voice of language echoing voice
    of nature, 109
  See also *Água Viva*; Cixous,
    Hélène; *Passion according to
    G. H., The*; Portuguese
    language
"Literature and the Right to Death"
  (Blanchot) (essay), 69, 74

Maldiney, Henri, 36
Mandelstam, Osip, 84, 109
Marcuse, Herbert, *One Dimensional
    Man*, 17, 118
Marx, Karl, 18
meanwhile, the, time and, 56, 59
memory, exile of, 22–30
  Alzheimer's disease and, 27
  Aristotle on, 23
  bound to past, 22–23
  how of memory, 27
  as image of the lost, 23–24
  imagination and, 25

the leaving behind and, 29
photisms and, 28–29
present and, 25–26
Proust and Freud's meditations
  on, 26, 27
as "shimmering go-between," 28,
  29
Memory and Recollection (Aristotle),
  24, 52
Metaphysical Foundations of Logic
  (1928), 50
Metaphysics (Aristotle), 7
metaphysics, overcoming, 55–56
modernity, 3
movement, excess of, 18–19

Nachheit, defined, 21
Nancy, Jean-Luc, 16
Near to the Wild Heart (Lispector)
  (novel), 86–87, 109
neuter, language of, 106–8
  writing in gerundive time,
    108–11
Nietzsche, lightning example and
  being, 44
Nobokov, Vladimir, on *how* of
  memory, 27–28, 29
Novalis, Schriften
  on farsightedness (*Fernsichtigkeit*), 25
  on philosophy as homesickness, 3

Odyssey (Homer), 2
One Dimensional Man (Marcuse),
  17, 118

passages, French meaning as
  "stepping," 70
Passion according to G. H., The
  (Lispector) (novel)
  attempt to read remains a failure,
    101

dashes at beginning and end of, 98, 99–100

fear is of to be-being—the cockroach, 90–91

gerundive mode of time and, 12

how to read, 88, 89

interpretation of, 91–93

plot of, 89, 90

sentimentação term explained, 96

structure of hints at echo-graphy, 109

what happens in this reading, 101–3

written in neuter, 85

Peri fugés, De exilio, On Exile (Plutarch), 2

127, 2

Pessoa, Fernando, *Book of Disquiet*, 8, 122–23

phenomenology, 2, 94–95, 100

Phenomenology of the Spirit (Hagel), 100

photisms, memory, exile of, 28–29

Plato, 2, 23, 25, 78–79

Plutarch, *Peri fugés, De exilio, On Exile*, 2

Poem of the End (Tsvetaeva), 123

Pontévia, Jean-Marie, 65, 124–25

Portrait of the Artist as a Young Man (Joyce), 87

Portuguese language, 5–6, 9–10, 17, 85, 97–99, 108–9

postexistential condition, exile as, 15–30

    overview, 11–12

    afterness of existence, 19–22

    excess, times of, 15–19

    memory and, 22–30

postmodernity, 3–4

presentism, 123

Proust, Françoise, on memory, 17, 26

"Question Concerning Technology" (Heidegger) (lecture), 48

Quignard, Pascal, 6

reading time and time of reading, 83–97

"retrotopia" (backwards longing), 119

Richter, Gerhard, *Afterness*, 21–22

Rosa, Guimarães, 104

Rosenzweig, Franz, 3

Said, Edward, on memory, 28

Schelling, J. W., 2, 109, 116

Science of Logic (Hagel), 100

Seneca, on human soul, 2

Socrates, Heidegger on, 57–58

Space of Literature, The (Kafka), 71–72, 79

St. Augustine, on memory, 26

Step not Beyond, The (Blanchot)

    beginning passage from, 68

    double meaning of, 67

    fragmentary writing and, 62–63

    key passage from, 80–81

    step motif and, 62

    temporality of the between and, 61

tension of the imminent, 66–67

thesis of book, 7

Third of May, The (Goya) (painting), 78

Thomas the Obscure (Blanchot) (novel), 67

Time and Being (Heidegger) (lecture), 47

"tired," defined, 15–16

truth, Heidegger's concept of, 59

Tsvetaeva, Marina, 114–16, 123, 124

Uncanny, The (*Das Unheimliche*)
    (Freud), 116–17, 119

Vorhandenheit, 42

Wahl, Jean, 40
Weber, Samuel, 6
What Is Called Thinking?
    (Heidegger) (lecture series
    1952), 57

What Is That Philosophy?
    (Heidegger), 52–53
withdrawal, question of, 12, 59
Writing of the Disaster, The
    (Blanchot), 69–70
    exile and fragmentary belong
        together, 62
    fragmentary writing defined, 66,
        67
    repetitive energy and, 69–70